Modern Blacksmithing
Rational Horse Shoeing and Wagon Making

by John G. Holstrom

with an introduction by Roger Chambers

Self Reliance Books

Get more historic titles on animal and stock breeding, gardening and old fashioned skills by visiting us at:

http://selfreliancebooks.blogspot.com/

Introduction

I am pleased to present yet another title on Homesteading and Farm Life.

This volume is entitled "Modern Blacksmithing" and was published in 1904.

The work is in the Public Domain and is re-printed here in accordance with Federal Laws.

As with all reprinted books of this age that are intended to perfectly reproduce the original edition, considerable pains and effort had to be undertaken to correct fading and sometimes outright damage to existing proofs of this title. At times, this task is quite monumental, requiring an almost total "rebuilding" of some pages from digital proofs of multiple copies. Despite this, imperfections still sometimes exist in the final proof and may detract from the visual appearance of the text.

I hope you enjoy reading this book as much as I enjoyed making it available to readers again.

Roger Chambers

Fig. 11—C. E. COLBURN'S FARM AND STOCK BARN

1

Fig. 19—MR. LAWSON VALENTINE'S BARN, "HOUGHTON FARM," MOUNTAINVILLE, N. Y.

ILLUSTRATIONS.

ILLUSTRATIONS.

PREFACE

WHAT prompted the author to prepare this book was the oft‑repeated question, by blacksmiths and mechanics of all kinds, as well as farmers: "Is there a book treating on this or that?" etc., etc. To all these queries I was compelled to answer in the negative, for it is a fact that from the time of Cain, the first mechanic, there has never been a book written by a practical blacksmith on subjects belonging to his trade. If, therefore, there has ever been such a thing as "filling a long‑felt want," this must certainly be a case of that kind.

In medicine we find a wide difference of opinion, even amongst practitioners of the same school, in treating diseases. Now, if this is so where there is a system, and authority for the profession, how much more so must there be a difference of opinion in a trade where every practitioner is his own authority. I shall, therefore, ask the older members of the blacksmith fraternity to be lenient in their judgment if my ideas don't coincide with theirs. To the apprentice

and journeyman I would say: do as I do until you find
a better way.

The author has been eminently successful in his
practice, and his ideas have been sought by others
wherever he has been, blacksmiths coming even from
other States to learn his ways.

This little book is fresh from the anvil, the author
taking notes during the day while at work, compiling
the same into articles at night.

He is indebted to a number of writers for article, in
this book treating on subjects belonging to their
trades, in which they have been regarded as expert.

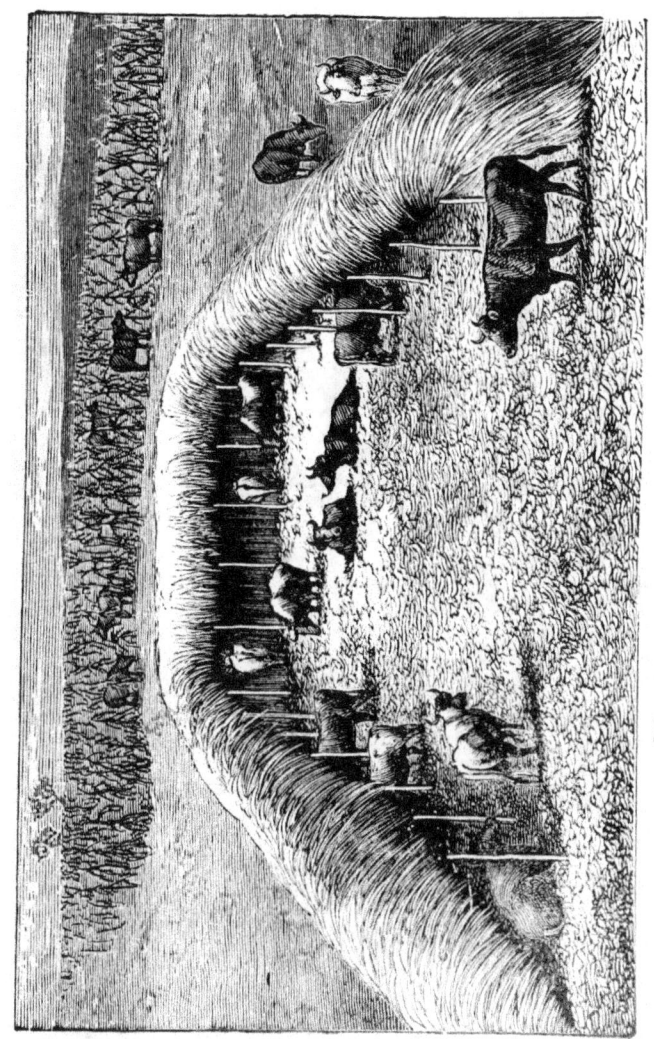

Fig. 123—STRAW SHELTER FOR CATTLE

Fig. 139—MR. GEORGE GRANT'S SHEEP CORRAL, VICTORIA, KAN.

8

*Now there was no smith found in all the land of Israel.—
1 Sam. 13:19.*

CHAPTER I

THE SMITH

FOR centuries the blacksmith has been a prominent person, and it is natural he should have been, when we consider the variety of work he had to do. From the heavy axle and tire, down to the smallest rivet in the wagon, they were all made by the smith. Bells and bits as well as the ornamental parts of the harness, they were all made by the smith. From the crowbar and spade down to the butcher and pocket knife, they were all made by the smith. The carpenter's tools, from the broadax and adz down to the divider and carving steel, they were all made by the smith. From the heavy irons in the fireplace down to the frying-pan and locks on the kitchen doors; knives and forks on the dining-table, they were all made by the smith. From the gun on the shoulder of the soldier and the saber in the hands of the officer, the spurs and pistol for the commander, they were all made by the smith. From the heavy anchor and its

9

chain to the smallest pulley in the rigging of the ship, they were all made by the smith.

From the weather vane on the church spire, and

THE SMITH

the clock in the tower down to the lock of the doors and the artistic iron cross over the graves in the church yard, they were all made by the smith. No wonder, then, that the smith was respected. Vulgar people swear by the devil, religious by the saints, but the Swedes (the makers of the best iron) prefer to swear by the smith. The smith was a well-liked person in society, respected and even admired for his skill, his gentlemanly behavior and good language. His stories and wit were the sole entertainment in many a social gathering. Things have changed in the last few decades. Most of the articles formerly made by the smith are now manufactured by machinery, and the respect for the smith is diminished in the same proportion. Not because there is not enough of the trade left to command respect—there is yet more left than any man can successfully learn in a short lifetime. But it has made it possible for men with less training and ability to enter the trade and consequently lower the standing of the smith. The result is, that there is a complaint that the smith is not esteemed as formerly, and I have been inclined to join in the lamentation. But instead of doing this I shall ask my brother smiths to unite with me in an effort to elevate the craft.

THERE ARE SMITHS AND SMITHS

I have had the pleasure of becoming acquainted with a great number of intelligent and respected smiths. People that did not know them would ask: "What is he?" and when informed that he is a blacksmith would say: "He doesn't look it; I thought he was a business man"; another, "He looks like a lawyer or a minister." From this you will understand how, in many cases, the blacksmith looks. A great preacher was announced to preach in a neighboring town, and ·I went to hear him. Just as I sat down in the pew one of the local smiths walked up to me and sat down by my side. He was a *blacksmith* and he "looked it." Under his eyes was a half moon in black; on both sides of his nose was a black stripe that had been there since his first day in the shop. His ears, well, you have seen a clogged-up tuyer iron. His clothes were shabby and his breath a strong mixture of tobacco and whisky, which made wrinkles on the nose of the lady in front of us. I was somewhat embarrassed, but the sermon began. As the congregation arose, I opened the hymnbook and my brother smith joined, and with a hand that looked like the paw of a black bear, he took hold of the book.

After service I was invited by the smith to dinner. Between a number of empty beer kegs we managed to reach the door of the house and everything inside looked the color of his trade. I looked around for books and other articles of culture and found a hand organ and a pack of cards. The only book or reading matter to be found was a weekly of the kind that tells of prize fights, train robberies and murder. I had a fair dinner and told my host that I had to start for home. By this time I was sick of his language—profanity, mixed with a few other words—and I started to leave. On my way to the livery stable I passed my friend's shop, and he said it would not be fair to leave before I had seen his shop. "I have," said he, "a very good shop." The shop was a building of rough boards 18x20—the average farmer has a better wood shed. A big wood block like the chopping block in a butcher shop, was placed so close to the forge that he could only get edgewise between. On this block was to be found, anvil and all his tools, the latter were few and primitive, and would have been an honor to our father Caiñ, the first mechanic and blacksmith. What thinkest thou, my brother smith? Having spent years to learn the trade you must submit to a comparison with smiths of this caliber. Their work being inferior they must work cheap, and in some, perhaps many, cases set the price on your work. Smiths of this kind cannot expect to be respected. There might be some show for them in Dawson City or among the natives in that vicinity, but not in civilized America.

INTEMPERANCE

NE of the chief reasons why the blacksmith is not so successful nor respected as before is his intemperance. The danger for the smith becoming a drunkard is g r e a t e r than for any other mechanic. It is often the case that when a customer pays a bill the smith is requested to treat. This is a bad habit and quite a tax on the smith. Just think of it—fifteen cents a day spent for liquor, will, in twenty-five years, amount to $9,000. Then add to this fifteen cents a day for cigars, which will, in twenty-five years, amount to $9,000 at ten per cent compound interest. If these two items would be saved, it will give a man a farm worth $18,000 in twenty-five years. How many smiths are there who ever think of this? I would advise every one to put aside just as much as he spends for liquor and tobacco; that is, when you buy cigars or tobacco for twenty-five cents put aside as much. When you buy liquor for one dollar put aside one dollar. Try this for one year and it will stimulate to continual effort in that direction. The best thing to do is to "swear off" at once, and if you must have it, take it out of business hours. Politely inform your friends that you must stop, or it will ruin you. If you drink with one you must drink with another, and the opportunity comes too often. When you have finished some difficult work you are to be treated; when you trust you are to be treated; when you accommodate one before another you are to be treated; when you order the stock from the traveling man you are to be treated. Some smiths keep a bottle in a corner to draw customers by; others tap a keg of beer every Saturday for the same purpose. No smith will ever gain anything by this bad practice. He will only get undesirable customers, and strictly temperance people will shun him for it. What he gains on one side he will lose on another. Besides this he will in the long run ruin himself physically and financially. Let the old smith quit and the apprentice never begin this dangerous habit. A smith that is drunk or half drunk cannot do his duty to his customers, and they know it, and prefer to patronize a sober smith.

RELIGION

TRUE religion is also an uplifting factor, and must, if accepted, elevate the man. I cannot too strongly emphasize this truth. Every smith should connect himself with some branch of the church and be punctual in attendance to the same. There is a great deal of difference between families that enjoy the Christianizing, civilizing and uplifting influence of the church and those outside of these influences. The smith outside of the church, or he who is not a member thereof will, in many cases, be found on Sundays in his shop or loafing about in his everyday clothes, his wife and children very much like him. The church member—his wife and children, are different. Sunday is a great day to them. The smith puts on his best clothes, wife and children the same. Everything in and about the house has a holiday appearance and the effect on them of good music and singing, eloquent preaching, and the meeting of friends is manifested in their language, in their lofty aims, and benevolent acts. Sunday is rest and strength to them.

Brother smiths, six days a week are enough for work. Keep the Sabbath and you will live longer and better.

INCOMPETENCY

Another reason the smith of to-day is not respected is his incompetency.

When a young man has worked a few months in a shop, he will succeed in welding a toe calk on a horse-shoe that sometimes will stay, and at once he begins to think he knows it all. There will always be some fool ready to flatter him, and the young man believes that he is now competent to start on his own hook. The result is, he hangs out his shingle, begins to practice horse-shoeing and general blacksmithing, and he knows nothing about either. Let me state here that horse-shoeing is a trade by itself, and so is black-smithing. In the large cities there are blacksmiths who know nothing about horse-shoeing, as well as horse-shoers who know nothing about blacksmithing, except welding on toe calks, and in many instances even that is very poorly done. In small places it is different. There the blacksmith is both blacksmith and horse-shoer. Sometimes you will find a black-smith that is a good horse-shoer, but you will never find a horse-shoer that is a good blacksmith. This is not generally understood. To many blacksmithing seems to mean only horse-shoeing, and our trade journals are not much better posted. Whenever a blacksmith is alluded to, or pictured you will always find a horse-shoe in connection with it. Yet there are thousands of blacksmiths that never made a horse-shoe in all their lives. Horse-shoeing has developed to be quite a trade, and if a man can learn it in a few years he will do well. I would not advise any young man to start out for himself with less than three or four years' experience. Every horse-shoer should make an effort to learn blacksmithing. He will be expected to know it, people don't know the difference; besides this, it will, in smaller cities, be hard to succeed with horse-shoeing alone. On the other hand, every blacksmith should learn horse-shoeing, for the same reasons. Therefore, seven or even ten years is a short time to learn it in. But, who has patience and good sense enough to persevere for such a course, in our times, when everybody wants to get to the front at once? Let every young man remember that the reputation you get in the start will stick to you. Therefore be careful not to start before you know your business, and the years spent in learning it will not be lost, but a foundation for your success. Remember, that if a thing is not worth being well done it is not worth being done at all. It is better to be a first-class bootblack or chimney sweep, than be a third-class of anything else.

Don't be satisfied by simply being able to do the work so as to pass, let it be first class. Thousands of mechanics are turning out work just as others are doing it, but you should not be satisfied to do the work as others are doing it, but do it right.

A MODERN GUILD

The blacksmiths and horse-shoers have at last put the thinking cap on, for the purpose of bettering their condition., So far nothing has been accomplished, but I am sure it will, in the long run, if they only keep at it. We are now living in the license craze age. From the saloon keeper down to the street peddler, they all howl for license, and unreasonable as it is, thousands of sensible men will cling to it in hopes that it will help.

We are, more or less, one-idea men, with fads and whims. Nations and organizations are just like individuals, ready to fall into a craze and we see it often. It is natural when we consider that nations and organizations are simple *one* man repeated so many times.

Simply look at the hero-worshiping craze went through at the close of the Spanish war. First, Lieutenant Hobson was the idol, and great was he, far off in Cuba. But, coming home, he made himself obnoxious on a tour through the country, and the worshipers were ashamed of their idol, as well as of themselves. Admiral Dewey was the next hero to be idolized, and he, too, was found wanting.

Physicians have their favorite prescriptions, ministers their favorite sermons. Politicians have their tariff and free trade whims, their gold or silver craze. Mechanics have their one ideal way of doing their work. I know horse-shoers that have such faith in bar shoes that they believe it will cure everything from contraction to heaves. Others have such a faith in toe weight that they will guarantee that in a horse shod this way the front quarters will run so fast that they must put wheels under the hind feet to enable them to keep up with the front feet; and in a three-mile race the front quarters will reach the stables in time to feed on a peck of oats before the hind quarters catch up.

In some States there is a union craze. All that these schemes will do is to prepare the legislatures for the legislation that will some day be asked of them. Unions have been organized and the objections are the same. I object to all these schemes because they fall short of their purpose.

Two years ago the horse-shoers of Minnesota asked the legislature to give them a license law. I wrote to a prominent member of the house of representatives and asked him to put his influence against the measure. He did so, with the result that the bill was killed so far as the counties and smaller towns were concerned. Such a law will only provide for an extra tax on the poor smiths and horse-shoers, and his chances of making a living will not be bettered, because no one will be shut out, no matter how incompetent.

TAXATION WILL NEVER RAISE THE STANDARD OF A MECHANIC

It deprives him of the means whereby to raise himself. Such a
law will only create offices to grease the machinery
for the political party in power.

THE only thing that will ever elevate the standard of workmanship is education, education and nothing but education. Give us a law that will provide for a certain degree of education before a boy is allowed to serve as an apprentice; and that he will not be allowed to start out for himself until he has served the full term, both as an apprentice and journeyman. And if intemperate, no diploma shall be issued to him. I see now that I was right when I opposed this law. The horse-shoers of Minnesota are now kicking and cursing the examining board. The National Convention of horse-shoers which was held in Cincinnati passed resolutions which were ordered transmitted to the governor of Illinois, requesting that the board of examiners now authorized to grant licenses to horse-shoers in that State, be changed, as "The board has failed to accomplish the purpose for which it was instituted—the elevating of the standard of workmanship of horse-shoers of that State." Unions are all right in every place where there is only one smith, let that smith unite with himself to charge a living price for his work and he is all right. Where there are more than one smith unions will only help the dishonest fellow. Such unions live but for a short time and then the smiths knife each other worse than ever.

In hard times (and hard times are now like the poor, "always with us,") a lot of tinkers start in the shoeing and blacksmith business. If they could make a dollar a day in something else they would stay out, but this being impossible, they think it better to try at the anvil. For them to get anything to do without cutting prices is out of the question, and so the cutting business begins, and ends when the regular smith has come down to the tinker's price. To remedy this we must go to the root of the evil. First, political agitation against a system whereby labor is debased.

This is a fact, in spite of all prosperity howling. Whenever there is trouble between labor and capital we will always find the whole machinery of the government ready to protect capital. The laboring men will not even be allowed to meet, but will be dispersed like so many dogs. They are the mob! But the capitalists, they are gentlemen! When the government wants a tailor for instructor in our Indian schools, or a blacksmith for the reservation, they get about

$600.00 per year. But, when a ward-heeler wants office he must have $5,000 per year. What induce-ment is it, under such conditions, for a young man to learn a trade? Laboring men, wake up!

But, as this will bring us into politics I shall leave this side of the question, for it would do no good. Thomas Jefferson, in the Declaration of Independence said: "Mankind are more disposed to suffer, while the evils are sufferable, than to right themselves by abolishing the forms to which they are accustomed." The laboring people will, in my judgment, suffer quite a while yet. In the meantime let us build up a fra-ternity on the ruins of the ancient guilds. Between the twelfth and the fifteenth centuries mechanics of all kinds prospered as never before, nor have they done it since. The reason for this was not a high protective tariff, or anything in that line, but simply the fruit of the guilds and the privilege they enjoyed from the state.

What we now need is a modern guild. I anticipate there would be some difficulty in securing the legis-lation necessary, but we will not ask more than the doctors now have. I cannot now go into detail; that would take more room and time than I can spare in this book.

NE thing is certain, we have a hard row to hoe, because, this is a government of injunctions, and any law on the statute book is in danger of being declared unconstitutional, according to the biddings of the money power, or the whim of the judges. One tyrant is bad, but many are worse.

I am no prophet, but will judge the future from the past. History will repeat itself, and Christ's teachings will be found true: "A house divided against itself cannot stand."

I will say so much, however, that no man should be allowed to start out for himself before he has served three years as an apprentice and two or three years as a journeyman. This should be proved by a certificate from the master for whom he has worked. This certificate to be sworn to by his master, one uninterested master and himself. No apprentice to be accepted without a certificate from the school superintendent that he has a certain knowledge in language and arithmetic and other branches as may be required. It shall not be enough to have worked a few days each year, but the whole time. With these papers he shall appear before three commissioners, elected by the fraternity and appointed by the governor of the State.

He shall pay not less than ten and not more than twenty-five dollars for his diploma. All complaint shall be submitted to these commissioners, and they shall have full power to act. If a practitioner acts unbecoming, runs down his competitor, charges prices below the price fixed by the fraternity, or defrauds his customers, such shall be reported to the commissioners, and, if they see fit, they can repeal or call in -his diploma and he shall not be allowed to practice in the State. These are a few hints on the nature of the modern guild we ought to establish. The fraternity should have a journal edited by one editor on literature and one on mechanics, the editor on mechanics to be a practical blacksmith with not less than fifteen years' experience. The editors are to be elected by the fraternity. This is all possible if we can get the legislation that the doctors have in many States. And why not?

Mechanics of to-day have a vague and abstract idea of what is meant by journeyman and apprenticeship. In Europe there is yet a shadow left of the guilds where these were in existence.

When I learned my trade I worked some time with my father in Sweden, then I went over to Norway and worked as an apprentice in Mathison & Johnson's machine, file and lock factory of Christiania. I was requested to sign a contract for four years. In this contract was set forth the wages I was to receive, and what I was to learn each year. Everything was specified so that there would be no room for misunderstanding. The first two weeks I worked, they simply drilled me. I was given a good file and a piece of iron, this iron I filed square, round, triangle, hexagon and octagon I wore out files and pieces of iron one after another, the master giving instructions how to stand, hold the file, about the pressure and strokes of same, etc. The same careful instructions were given in blacksmithing. The apprentice was given some work, and he had to forge it out himself, no matter what time it took, nor did it make any difference if the job, when done, was of any use, the apprentice was simply practicing and accustoming himself to the use of tools. Thus the elementary rules were learned in a few weeks, and the apprentice made capable of doing useful service that would repay for the time lost in the start.

LITERATURE

AVING thoroughly learned the trade, it is important to keep posted in this matter by reading books and trade journals. As far as books are concerned, we have a few treating on horse-shoeing, with both good and bad ideas. As to blacksmithing, this book, ' Modern Blacksmithing,'' is the first in that line, written by a practical blacksmith and horse-shoer.

Our trade journals must be read with discrimination. They are mostly edited by men having no practical experience in the trade, and are therefore not responsible for the articles these papers contain. Many articles are entirely misleading. Blacksmiths having more experience with the pen than the hammer, and anxious to have their names appear in print, write for these journals.

Prize articles are also doing more harm than good, the judges giving the prizes to men with ideas like their own, not being broad-minded enough to consider anything they don't practice themselves, and the result is a premium on old and foolish ideas.

But we should not stop at this. We should read much. Anything, except bloody novels, will help to elevate the man. No smith should think it idle to **read** and study. "Every kind of knowledge," observes a writer, "comes into play some time or other, not only systematic study, but fragmentary, even the odds and ends, the merest rag-tags of information." Some fact, or experience, and sometimes an anecdote, recur to the mind, by the power of association, just in the right time and place. A carpenter was observed to be very particular and painstaking in repairing an old chair of a magistrate, and when asked why, said: "I want this chair to be easy for me to sit in some time." He lived long enough to sit in it.

Hugh Miller found time while pursuing the trade of a stone mason, not only to read, but to write, cultivating his style till he became one of the most facile and brilliant authors of the day. Elihu Burritt acquired a mastery of eighteen languages and twenty-two dialects, not by rare genius, which he disclaimed, but by improving the bits and fragments of time which he had to spare from his occupation as a blacksmith.

Let it be a practice or a habit, if you will, to buy at least one book every year, and to read the same, once, twice, thrice, or until its contents are indelibly impressed upon your mind. It will come back to your mind and be useful when you expect it the least.

CHAPTER II

O other mechanic will try to turn out such a variety of work with so few tools as the blacksmith, even when the smith has all the tools to be had, he has few in proportion to the work. There are a class of smiths who will be content with almost nothing. These men can tell all about the different kinds of tobacco; they can tell one kind of beer from another in the first sip, and the smell of the whisky bottle is enough for them to decide the character of the contents, but when it comes to tools which belong to their trade, they are not in it. It ought to be a practice with every smith to add some new tool every year. But if they are approached on the subject they will generally say, "Oh, I can get along without that." With them it is not a question of what they need, but what they can get along without.

Some smiths have the Chinaman's nature (stubborn conservatism) to the extent that they will have nothing new, no matter how superior to their old and inferior tools; what they have been used to is the best.

When the hoof shears were a new thing I ordered a pair and handed them to my horse-shoer, he tried them for a few minutes and then threw them on the floor and said, "Yankee humbug." I picked them up and

29

tried them myself, and it took a few days before I got used to them, but then I found that they were a great improvement over the toe knife. I told my horse-shoer tc use them and after a while he could not get along without them, but would yet have used his toe knife if it had not been for the fact that he was com-pelled to use them. If it was not for the conservatism by which we are all infected more or less, we would be far more advanced in everything.

The mechanic that has poor tools will in every case be left behind in competition with the man with good tools in proper shape. There are smiths who will take in all kinds of shows and entertainments within fifty miles, but when it comes to tools, oh, how stingy and saving they are. There is no investment which will bring such a good return as first-class tools do to a mechanic. The old maxim, "A mechanic is known by the tools he uses," is true. Many of the tools used in the shop can be made by the smith. If less time is spent in the stores and saloon there will be more time for making tools.

I shall, in this chapter, give a few pointers how to make some of the tools used. I will not spend any time in explanation about the more intricate tools like drill presses and tools of that kind, because no smith has experience or facilities to make tools of this char-acter that will be worth anything. I shall simply give a few hints on the most common tools used, with illus-trations that will be a help to new beginners. Before we go any further let me remind you of the golden rule of the mechanic, "A place for everything and everything in its place." Some shops look like a scrap iron shed, the tools strewn all over, and one-tenth of the time is spent in hunting for them. I shall first say a few words about the shop and give a plan. This plan is not meant to be followed minutely, but is simply a hint in that direction.

THE SHOP

In building a shop care should be taken in making it convenient and healthy. Most of the shops are built with a Ligh floor. This is very inconvenient when machinery of any kind is taken in for repairs, as well as in taking in a team for shoeing. Around the forge there should be a gravel floor. A plank floor is a great nuisance around the anvil. Every piece cut off hot is to be hunted up and picked up or it will set fire to it. I know there will be some objection to this kind of floor but if you once learn how to keep it you will change your mind. To make this floor take sand and clay with fine gravel, mix with coal dust and place a layer where wanted about four inches thick. This floor, when a little old, will be as hard as iron, provided you sprinkle it every night with water. The dust and soot from the shop will, in time, settle in with it and it will be smooth and hard. It will not catch fire; no cracks for small tools or bolts to fall through; it will not crack like cement or brick floors. If your shop is large then make a platform at each end, and a gravel floor in the center, or at one side, as in figure

1. This floor is cool in summer and warm in winter, as there can be no draft. The shop should have plenty of light, skylights if possible. The soot and dust will, in a short time, make the lightest shop dark. The shop should be whitewashed once a year. Have

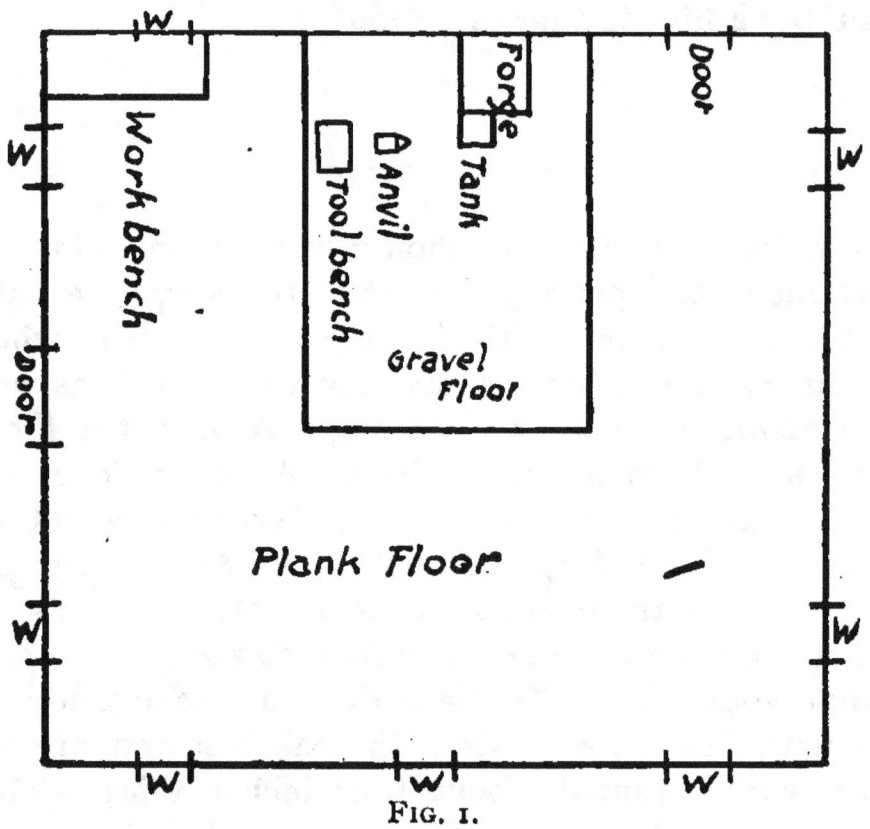

FIG. 1.

plenty of ventilation. Make it one story only if convenient to do so, as an upper story in a blacksmith shop is of very little use. The shop is the place where the smith spends most of his time and he should take just as much care in building it, as a sensible house-keeper does in the construction of her kitchen.

THE FORGE

The forge can be made either single or double, square or round. The square is the best as it can be placed up against the wall, and you will then have more room in front of it. The round forge will take more room, if it is placed in the center of the floor there will be no room of any amount on any side and when the doors are open the wind will blow the fire, cinders and smoke into the face of the smith. This is very uncomfortable. The smokestack, if hung over the fire will sometimes be in the way. Of course the hood can be made in halves and one half swung to the side, but it will sometimes be in the way anyhow, and it seldom has any suction to carry away the smoke and cinders.

THE ANVIL

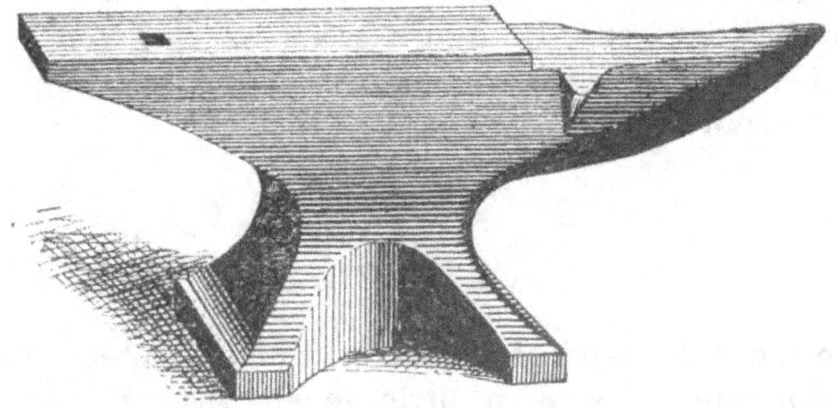

The anvil should not be too close to the forge, as is often the case in small country shops. Make it six feet from center of fire to center of anvil. The anvil should not be placed on a butcher block with the tools on, but on a timber the same size as the foot of the anvil. Set the timber down in the ground at least three feet. For heavy work the anvil should stand low in order to be able to come down on it with both hammer and sledge with force. When the smith has his hands closed the knuckles of his fingers should touch the face of the anvil and it will be the right height for all-around blacksmithing.

COAL BOX

Close to the forge under the water tank or barrel should be a coal box 18 x 24 x 16 inches, this box to be dug down in the ground and so placed that one end will protrude from under the barrel or tank far enough to let a shovel in. This opening can be closed with a lid if the tools are liable to fall into it. In this box keep the coal wet. In figure 1 a plan is given from which you can get an idea of a shop and how to place the tools and different articles needed.

TOOL TABLES

On the right hand of the anvil should be a tool bench or tool table 20 x 20, a little lower than the anvil. Outside, on three sides and level with the table, make a railing of 1¼ inch iron, about 1½ inch space between the table and railing; this makes a handy place for tools and near by.   Many blacksmiths have no other place than the floor for their tools, but there is no more sense in that than it would be for a carpenter to throw his tools down on the floor all around him. There ought to be "a place for every tool and every tool in its place."

TOOL TABLE.

THE HAMMER

When a lawyer or a minister makes his maiden speech he will always be in a great hurry on account of his excitement.   The sentences are cut shorter, broken, and the words are sometimes only half pronounced.   After a few years' practice he will be more self-possessed and the speech will be changed from unintelligible phrases to logical oratory. When the carpenter's apprentice first begins to use the saw, he will act the same way—be in a great hurry—he will run the saw at the speed of a scroll saw, but only a few inches of stroke; after some instructions and a few year's practice the saw will be run up and down steady and with strokes the whole length of the blade. When the blacksmith's apprentice begins to use the hammer he acts very much the same way. He will press his elbows against his ribs; lift the hammer only a few inches from the anvil and peck away at the speed of a trip hammer. This will, in most cases, be different in a few years. He will drop the bundle— that is, his elbows will part company with his ribs, the hammer will look over his head, there will be full strokes and regular time, every blow as good as a dozen of his first ones. Some smiths have the foolish habit of beating on the anvil empty with the hammer, they will strike a few blows on the iron, then a couple of blind beats on the anvil, and so on. This habit has been imported from Europe, free of duty, and that must be the reason why so many blacksmiths enjoy this luxury.

THE SLEDGE

In Europe great importance is laid upon the position taken by the apprentice and the manner he holds the sledge. The sledge is held so that the end of it will be under his right armpit, when the right hand is **next** to the sledge, and under his left arm when the left hand is nearer the sledge. In this unnatural position it is next to impossible to strike hard and do it for any time. This is another article imported free of duty, but few Americans have been foolish enough to use it. In this country the apprentice will be taught to use the tools in a proper way.

The end of the sledge-handle will be to one side; at the left, if the left hand is at the end of the handle, and at the right if the right hand is at the end of the handle; and be down between his feet when the handle's end must be low. The apprentice should stand directly in front of the anvil.

In swinging, the sledge should describe a circle from the anvil close down to the helper's feet and up over his head and down to the anvil; this is a perpendicular circle blow. Be sure not to give it a horizontal start; that is, with one hand close to the sledge the apprentice starts out either in the direction of the horn or the butt end of the anvil, and then up while both hands should clasp the extreme end of the handle close together the sledge should be dropped down to the feet then up. The hold taken should not be changed, but the hands held in the same place. (See figure 4.)

For ordinary use a nine-pound sledge is heavy enough, a large sledge will give a bump, while a small one will give a quick good blow, it is only occasionally and for special purposes a large sledge is needed, even an eight-pound sledge will do. Try it, and you will be surprised how nice it works.

With these preliminary remarks we shall now begin
to make a few tools.   We will begin with the blacksmith's tongs.   I shall only give an idea how to forge
the jaws, and every man that needs to make them has

FIG. 4.

seen enough of this simple tool to know what kind is
needed, and what he has not seen will suggest itself
to every sensible smith.

BLACKSMITH'S TONGS

Take a piece of one-inch square Swede iron, hold the iron diagonally over the anvil, with your left hand a little toward the horn, the end of the iron to reach out over the outside edge of the anvil. Now strike so that the sledge and hammer will hit half face over the anvil and the other half of the sledge and hammer out-

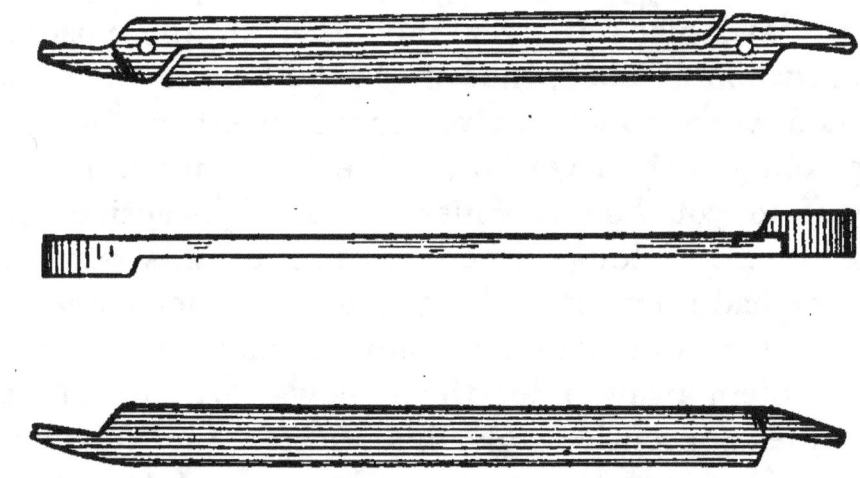

side of the anvil. Hammer it down to about three-eighths of an inch thick. Now pull the iron towards you straight across the anvil, give it one half turn toward yourself so that this side which was up, now will be towards yourself; the end that first was outside the anvil now to rest over the inner edge of the anvil, push the jaw up against the anvil until it rests against the shoulder made in the first move. Now hammer this down until it is the thickness of the jaw that is desired. Next, turn it over, with the bottom side up or the side that was down, up; push it out over the outside edge of the anvil again so far that the shoulder or set down you now have up, will be about an inch outside and over the edge of the anvil, now give a few blows to finish the jaw, then finish the shanks and weld in half inch round iron to the length desired. The jaws should be grooved with a fuller, if you have none of the size required take a piece of round iron and hammer it down in the jaws to make the groove. Tongs grooved this way will grip better. Next, punch a hole in one jaw, place it over the other in the position wanted when finished, then mark the hole in the other jaw, and when punched rivet them together, the jaws to be cold and the rivet hot. The following story will suggest to you how to finish it. An apprentice once made a pair of tongs when his master was out, and when he had them riveted together could not move the jaws. As he did not know how to make them work he laid them away under the bellows. At the supper table the apprentice told his master the following story: An apprentice once made a pair of tongs and when he had them riveted together he could not move the jaws, and as he did not know what to do he simply threw them away, thinking he must have made a mistake somehow. "What a fool," said the master, "Why didn't he heat them." At the next opportunity the apprentice put his tongs in the fire and when hot they could be worked very easily.

HOW TO MAKE A HAMMER

Take a piece of tool steel 1¼ inches square, neat it red hot. Now remember here it is that the trouble begins in handling tool steel. If, in the process, you ever get it more than red hot, it is spoiled, and no receipt, or handling or hammering will ever make it good again. The best thing in such a case is to cut off the burnt part in spite of all proposed cures. This must be remembered whenever you heat tool or spring steel. If the burnt part cannot be cut off, heat it to a low heat, cool it in lukewarm water half a dozen times, this will improve it some, if you can hammer it some do so. Now punch a hole about two inches from the end with a punch that will make a hole 1⅛ x ⅜. If the punch sticks in the hole, cool it off and put a little coal in the hole that will prevent the punch from sticking. This is a good thing to do whenever a deep hole is to be punched. Be sure that the hole is made true. Next, have a punch the exact size of the hole wanted when finished, drive it in and hammer the eye out until it has the thickness of about ⅜ of an inch on each side and has a circle form like No. 2, Figure 5.

In order to do this you may have to heat the eye many times, and upset over it with the punch in the eye. This done put in the bottom fuller and with the top fuller groove it down on each side of the eye, like the cut referred to. Now dress down the face then the peen-end. When finished harden it in this way: Heat the face-end first to a low red heat, dip in water about an inch and a half, brighten the face and watch for the color. When it begins to turn blue cool off but don't harden the eye. Wind a wet rag around the face end and heat the peen-end, temper the same way. With a piece of iron in the eye, both ends can be hardened at the same time, but this is more difficult, and I would not recommend it.

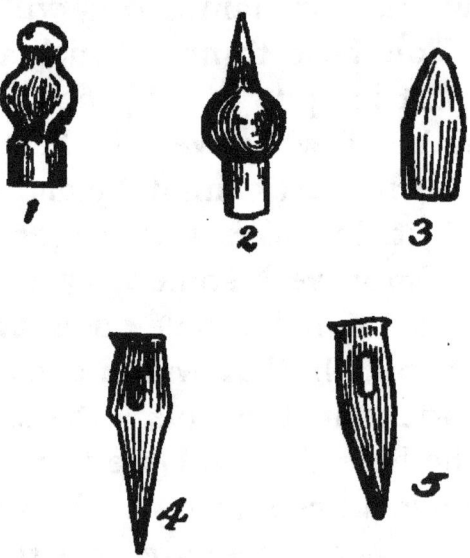

FIG. 5.

For ordinary blacksmithing a flat peen hammer is the thing, but I have seen good blacksmiths hang on to the machinist's hammer as the only thing. See No. 1, Figure 5. This hammer is more ornamental than useful in a blacksmith shop. The hammer should be of different sizes for different work, light for light work, and for drawing out plowshares alone the hammer should be heavy.

For an ordinary smith a hammer of two up to two and one-half pounds is right. Riveting hammers should be only one pound and less. No smith should ever use a hammer like No. 3, in Figure 5. This hammer I have not yet been able to find out what it is good for. Too short, too clumsy, too much friction in the air. I have christened it, and if you want my name for it call it Cain's hammer. It must surely look like the hammer used by him, if he had any.

HOW TO MAKE CHISELS

A chisel for hot cutting, see Figure 5, No. 4. This chisel is made of 1¼ square tool steel. Punch a hole 1⅛ x ¼ x ½ about three inches from the end, the eye should be narrow in order to leave material enough on the sides to give it strength. When eye is finished, forge down below it, not on the head-end, with top and bottom fullers, like cut. This gives the chisel a better shape. Now dress down the edge, then heat to a low cherry red, and harden, brighten it and when the color is brown cool off.

COLD CHISELS

Use same sized steel as above referred to, make it like No. 5, Figure 5. To distinguish it from the hot cutting chisel, and to give it more strength, in hardening this chisel, draw the temper until it is blue. This is the right temper for all kinds of cold chisels.

SET HAMMER

One might think that anybody knows how to make a set hammer, if every smith knows it, I don't know, but I do know that there are thousands of smiths who have never had a set hammer nor know its use. To make one: Take a piece of tool steel 1¼ x 1¼ inches, punch a hole about two inches from the end, the hole to be 1¼ x ⅜. Now cut off enough for head. Make the face perfectly square and level, with sharp corners, harden and cool off when the temper turns from brown to blue. This is a very important little tool and for cutting steel it is a good deal better than the chisel. Plow steel of every kind is easier cut with this hammer than any other way. In cutting with the set hammer hold the steel so that your inner side of the set hammer will be over the outside edge of the anvil. Let the helper strike on the outside corner of the set hammer and it will cut easy. The steel to be cut should be just a little hot, not enough to be noticed. If the steel is red hot the set hammer cannot cut it. The heat must be what is called blue heat. I would not be without the set hammer for money, and still I often meet smiths who have never seen this use made of the set hammer. Plow points, corn shovels, and seeder shovels are quicker cut with this tool than any other way, with the exception of shears.

TWIST DRILLS

Twist drills are not easy to make by hand, as they should be turned to be true, but a twist drill can be made this way.   Take a piece of tool steel round and the size of the chuck hole in your drill press.   Flatten it down to the size wanted, heat, put the shank in the vise, take with the tongs over the end and give one turn to the whole length, turn to the left.   When finished be sure that it is not thicker up than it is at point, and straight.   Now harden, heat to a low cherry red, cool off in luke-warm water—salt water, if you have it—brighten it and hold over a hot bar of iron to draw temper, cool off when brown, the whole length of the twist should be tempered.

Another way to make a drill is to just flatten the steel and shape to a diamond point and bend the shares forward. This is a simple but good idea and such drills cut easy.   In cooling for hardening turn the drill in the water so that the edge or shares are cooled in proportion to point, or the shares will be too soft and the point of such a drill too hard.   Our trade journals, in giving receipts for hardening drills, often get watch-makers receipts.   This is misleading: watch-makers heat their drills to a white heat.   Now, remember, as I have already said, when your drill or tool of this kind is heated to this heat the best thing to do is to cut that part off.   It is different with watch-makers, they do not look for strength, but hardness. They run their drills with a high speed, cut chips that cannot be discerned with the naked eye, and must

have a drill that is hard like a diamond.   For drilling
iron or steel the drill does not need to be so very hard,
but tough rather, because of the slow speed and thick
chips.   Few smiths have been able to master the sim-
plest tempering, and they think if they could get a
complicated receipt they would be all right.   We are
all more or less built that way.   Anything we do not

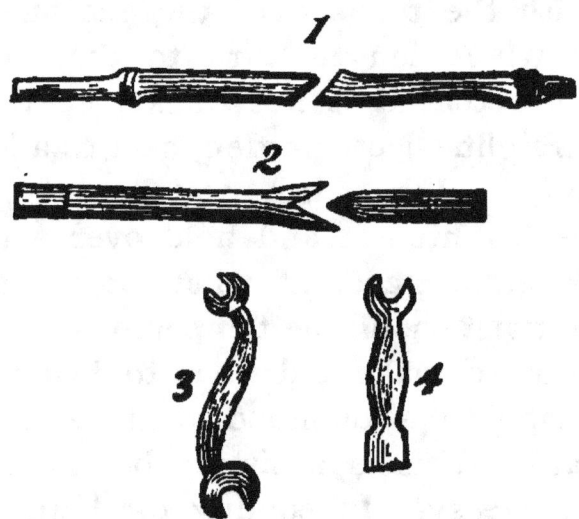

FIG. 6.

understand we admire. Simple soft water and the
right heat is, in most cases, the only thing needed for
hardening. I had occasion to consult a doctor once
who was noted for his simple remedies. A lady got
some medicine and she wanted to know what it was so
she could get it when the doctor was not at home, but
he refused to reveal it to her. When the lady had left
the doctor told me the reason why. "This lady," said
the doctor, "does not believe in simple remedies which
she knows, but believes in those remedies she knows
nothing about." I think it is better for us to try to

understand things and not believe much in them
before we understand them.

S WRENCH

See Figure 6, No. 3. This wrench is for ⅜ nut on
one end and ½ on the other, just the kind for plow
work. To make one, take a piece of tool steel 1½ x ⅝,
start as you see in No. 4, Figure 6. Set the jaws down
with the fullers, punch a round hole as in end No. 4,
cut out from hole and finish the jaws to make the right
length, now bend it in S shape and finish. This makes
the best wrench. Do not heat over a red heat.

ROCK DRILLS

Few blacksmiths know how to make a rock drill.
Take a piece of round or octagon steel, the desired
length and thickness, shape it, but it must be remembered that if during the process you ever get it over a
red heat there is no use to proceed, but just cut off
that much and start again, no hardening will prevail
if it is burnt. The trouble begins when you put the
steel into the fire, and you must watch until you have
it finished. When ready to harden heat it to a cherry
red heat, cool in water not too cold, brighten and
watch for temper. When it is yellow, cool it off, but
not entirely, take it out of the water before it is quite

cold and let it cool slowly, this will make the drill both hard and tough. By this simple process I have been able to dress drills and get such a good temper than only two per cent would break. Another way to harden is to heat to a very low heat and cool it off entirely at once. A third way is to temper as first stated and when yellow set the drill in water only one half an inch deep and let it cool. By this process a good per cent will break just at the water line.

"Be sure you are right then go ahead."—Davy Crocket.

CHAPTER III

HOW TO STRIKE AND TURN THE IRON—RULES FOR SMITH AND HELPER

THE smith should never turn the iron on the helper's blow, he should turn on his own blow, that is, never turn the iron so that the helper's blow will hit it first because he is not prepared for it and cannot strike with confidence, but the smith will not be bothered by turning the iron for himself as he knows when he turns and is prepared for it. The smith should strike the first blow in starting, or signal the helper where to strike, in case the smith cannot strike the first blow. The smith calls the helper by three blows on the anvil with his hammer, and when the smith wants the helper to cease striking he taps with the hammer twice on the anvil. The helper should strike the blow he has started when the smith signals him to stop. The helper should watch the time of the smith's hammer; if fast, keep time with it, if slow, keep time with it. The helper should strike where the smith strikes or over the center of the anvil. The helper should always lift the sledge high, in order to give the smith a chance to get in with the hammer.

49

THE FIRE

It is proper before we go any farther to say a few words about the fire.

An old foreman in the blacksmith department of a factory told me once in a conversation we had about

"CORRECT POSITION" AT THE ANVIL

the fire, that he had come to the conclusion that very few blacksmiths have learned how to make a good fire. It takes years of study and practice before the eye is able to discern a good fire from a bad one. A good fire must be a clear fire, the flame must be concentrated and of a white color. Even the nose must serve to decide a bad fire from a good one. A strong sulphur smell indicates a poor fire for welding. In order to get a good fire there must be, first, good coal; second, plenty of it. It is no use to pile a lot of coal on an old fire, full of cinders and slag. The fire-pot must be clean. Many blacksmiths are too saving about the coal. They take a shovel of coal, drop it on the forge in the vicinity of the fire and sprinkle a handful of it in the fire once in a while. In such a case it is impossible to do good work and turn it out quick. Have a scoop shovel and put on one or two shovels at a time, the coal should be wet. Then pack it in the fire as hard together as you can. Sprinkle the fire with water when it begins to spread. In this way you get a hard fire. The flames are concentrated and give great heat. Saving coal is just like saving feed to a horse, or grub to your apprentice. Neither will give you a good day's work unless he has all he wants to eat. The fire, of course, should be in proportion to the work, but in every case should the fire be large enough to raise it up from the tuyer iron as much as possible. In a small fire the blast strikes directly on the iron and it begins to scale off; in a good fire these scales melt and make it sticky, while in a low and poor fire the scales blacken and fall off. This never happens if the fire is full of good coal and high up from the tuyer iron.

Good strong blast is also necessary for heavy work. There is an old whim about the fire that everybody, farmers and others, as well as blacksmiths, are infected with, and that is, if a piece of brass is put in the fire it renders the fire useless to weld with. Now, while it is a fact that brass is not conducive to welding it takes a good deal of it before the fire is made useless. One smith will not dare to heat a galvanized pipe in his fire, for fear it will spoil it, while another smith will weld a piece of iron or steel to such a pipe without difficulty. Don't swear and curse if the fire is not what you expect it to be, but simply make it right. Some smiths have the habit of continually poking in the fire, if they weld a piece of iron they never give it rest enough to get hot, but turn it over from one side to another and try to fish up all the cinders and dust to be found in the fire. This is a bad habit. Yellow colored fire is a sign of sulphur in the fire and makes a poor fire for welding. Dead coal makes a poor fire.

TUYER IRON

One of the chief reasons for a poor fire is a poor blast. No patent tuyer will give blast enough unless you run it by steam and have a fan blower. Ninety per cent of the blast is lost in transmission through patent tuyers. The only way to get a good blast is to have a direct tuyer, and one with a water space in.

To make a direct tuyer take a pipe $1\frac{1}{4}$ x 12 inches long, weld around one end of this pipe an iron $3\frac{5}{8}$ to make it thick on the end that is in the fire, flare out the other end for the wind pipe to go in and place it horzontal in the fire and fill up around it with fireproof clay. This gives the best fire. The only objection to

this tuyer is that where soft coal is used, as is mostly
the case in country shops, it gets hot and clogs up,
but with a strong blast and good hard coal it never
gets hot, provided the fire is deep enough. From five
to eight inches is the right distance from the tuyer to
the face of the fire. In factories this kind of tuyer is
used, and I have seen them used for ten years, and

WATER TUYER

never found them to clog once. The tuyer was just as
good after ten years use as it was when put in.

To make a water tuyer take a pipe 1¼ x 12, weld a
flange on each end for water space, now weld another
pipe over this, and bore holes for ¼ inch pipes in the
end, where the blast goes in. One hole on the lower
or bottom side should be for the cold water to go in
through, and one hole on the upper side for the hot
water to go out through. These pipes to connect with
a little water tank for this purpose. The pipes should

be watched so that they will not be allowed to freeze or clog, as an explosion might follow. These tuyers never clog. I now use one that I have made as above described. The dealers now have them to sell. Any smith can get them as they are hard to make by the average smith.

BLOWERS

I have tried many kinds of blowers and I shall give my brother smiths the advantage of my experience.

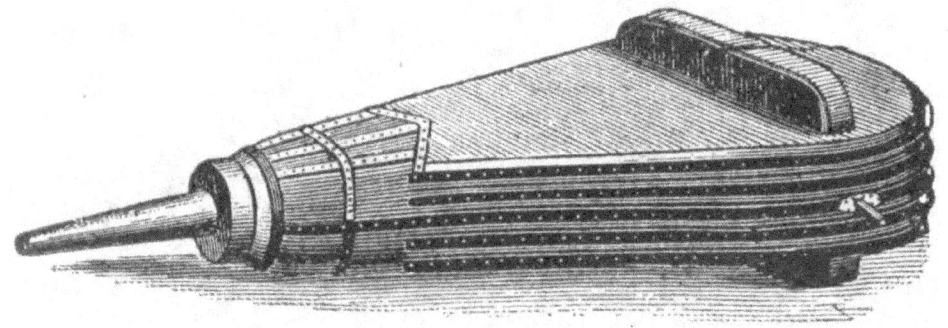

Portable forges run with fan blowers are fair blowers if you are strong enough to pump away at high speed, but it takes a horse to do that, and as soon as you drop the lever the blast ceases. Root's blower works easier, but the objection is the same, as soon as you drop the crank the blast stops. Besides this trouble, this blower is often in the way. I have never found anything to beat the bellows yet, if you only know how to use them.

Never take a set of bellows less than 48 extra long. Cut the snout off so that it will give a hole 1½, and with a water tuyer this blower cannot be beaten, except by a fan blower run by steam. The bellows should be hung over head to be out of the way. When these bellows are full of wind they will blow long enough after you have dropped the lever to do quite a good many things around the forge, and to handle the iron in the fire with both hands as is often necessary.

WELDING IRON

Welding iron is easy and no other welding compound is needed than sand, unless it is a case when the iron is liable to burn or scale off, borax will prevent this. There are three kinds of welds, butt, lap and split. The butt weld is most used in welding iron. The ends should be rounded off a little so that the center will weld first. Weld the ends this way either in the fire or on the anvil, butting the ends while you strike over and dress down the weld. In welding lap welds upset the ends and make them a good deal heavier than the size of the iron is; then lap the ends with a short lap. New beginners will always make a long lap. This is wrong, for if the lap is long it will reach beyond the upset part and the ends cannot then be welded down, without you make it weak. If soft steel is welded cut a short cut with the chisel in the center of the lap, as shown in Figure 6, No. 1. This cut will hook and prevent the ends from slipping; if properly prepared this weld will not show at all when done.

SPLIT WELDS

. Split weld is preferable when steel is to be welded, especially tool steel of a heavy nature, like drill bits for well drillers.

If the steel is welded to iron, split the iron and draw out the ends as thin as possible and make it the shape shown in Figure 6, No. 2. Taper the steel to fill the split made in the iron, when it fits perfectly cut beard in it to catch in the lips of the iron when fitted in. See Figure 6, No. 2. When finished heat the split end and cool off the tapered end. Place the tapered end snug up in the split and hammer it together with a heavy sledge. If there is any crack or opening at the end of the tapered end, plug it up with iron plugs, if this is not done, these holes will be almost as they are, because it is hard to weld a heavy shaft or drill, or rather, it is hard to hammer them together so the holes will close in. Now heat, but if you have tool steel go slow, or your steel will burn before the iron is hot enough. Weld the lips while the rod or drill is in the fire. For this purpose use a hammer with an iron handle in. When the lips are welded all around take it out and let two good helpers come down on it with all their might. When welded smooth it up with the hammer or flat hammer.

WELDING STEEL

Welding steel is quite a trick, especially tool or spring steel. The most important part to remember is, to have a good clean fire, and not to over heat the steel. To a good smith no other compound is needed than borax, but if this is not sufficient, take some borings from your drill, especially fine steel borings, and cover the weld with this and borax, and if a smith cannot weld with this compound there is no use for him to try. Most of the welding compounds are inferior to this, but some smiths would rather believe in something they don't know anything about; another will not believe in anything he can get for nothing.

BANDS OR HOOPS

When a round object is to be ironed or a hoop put on to anything round, measure, that is, take the diameter then multiply by three, add three times the thickness of the iron (not the width), add to this one time the thickness of the hoop for the weld and you have the exact length of the iron needed; in other words, three times the diameter, four times the thickness of the band. This is a simple rule, but I know a good many old smiths who never knew it.

SEEDER SHOVELS

To weld seeder shovels is no easy job. Prepare the shovel; shape almost to it proper shape, draw out the shanks, weld the points first, heat shovel and shank slow, then fit them together so that no cinders can get in between. Now remember, if your fire is not at least five inches up from the tuyer iron, and clear, it is no use to try. Hold your shovel in the fire, shank down. Heat slow, use borax freely and apply it on the face side of the shovel to prevent it from burning. When ready, weld it over the mandrill and the shovel will have the right shape. If soft center, harden like a plow lay.

DRILLING IRON

Every smith knows how to drill, sometimes it gives even an old smith trouble. The drill must be true, the center to be right, if one side of the drill is wider than the other or the drill not in proper shape the hole will not be true. For centuries oil has been used for drilling and millions of dollars have been spent in vain. It is a wonder how people will learn to use the wrong thing. I don't think that I have ever met a man yet who did not know that oil was used in drilling. In drilling hard steel, turpentine or kerosene is used as oil will then prevent cutting entirely. Nothing is better than water, but turpentine or kerosene is not as bad as oil; if you think water is too cheap use turpentine or kerosene. I had occasion once to do a little work for a man eighty years old, and when I drilled a hole, used water. The old man asked if water was as good as oil, and when informed that it was better, said: "I used to be quite a blacksmith myself, I am now eighty years old, too old to do anything, but I am not too old to learn." It ought to suggest itself to every smith that while oil is used in boxes to prevent cutting, it will also prevent cutting in drilling.

HOW TO DRILL CHILLED IRON

First prepare a drill which is thicker at the point than usual, and oval in form, then harden it as follows: heat to a low cherry red heat and cool in the following hardening compound: two quarts soft water, one-half ounce sal-ammoniac, salt, three ounces. Don't draw the temper, for if you have the right heat you will get the right temper. Now drill and use water, not oil. Feed carefully but so the drill will cut right along. If you have no chance to get the compound, harden in water but draw no temper, let it be as hard as it will.

If the iron is too hard to be drilled and you can heat the same do so, heat to a low red heat and place a piece of brimstone just where the hole is to be; this will soften the iron through, so the hole can be drilled. Let it cool slowly.

STANDING COULTERS

Standing coulters are made of different materials and of different shapes. Take a piece of iron $2\frac{1}{4}$ x $\frac{1}{2}$, twenty-eight inches long. Cut off the end after you have thinned it out about 5 inches from the end, cut diagonally   Now weld the cut-off piece to the main shank.   The cut-off piece to be laid on the outside and welded, bend the iron as soon as it is welded so that it has the shape of the coulter, draw out a good point and sharpen the iron just the same as if it was a finished coulter.   This done, cut off a piece of steel, an old plow lay that is not too much worn will do, cut

STANDING COULTER

the shape of the coulter you have now in the iron, and let the steel be half an inch wider than the iron, but on the point let it be as long as it will, because the point ought to be quite long, say about nine inches. Next draw the steel out thin on the upper end, heat the iron red hot, place it on the anvil outside up, put a pinch of borax on it at the heel, then a pinch of steel borings, place the steel on top of this and keep in position with a pair of tongs; now hold it on the fire heel down, and heat slow.   When it is hot let the helper strike a pressing blow or two on it and it will stick until you have taken the next weld.   Put borings and borax between steel and iron for each weld. When finished, the angle should be that of the square; that is, when you place the coulter in the square the shank should follow one end of the square and the foot of the coulter the other. The edge of the outside side should follow the square from the point up. When it does it looks like a hummock in the coulter but it is not. Old breakers will be particular about this as it will cut a clean furrow if it is made in this way and it will work easier. If the edge stands under the square the coulter will wedge the plow out of land and make a poor furrow. Next finish the chisel point, soft or hard steel as you please; weld it to the coulter on the inside, that is, the side next to the furrow.

Last punch or drill the hole in the heel. The coulter should not be hardened except a little on and along the point. There is no need of a double chisel point, such a point will be too clumsy and run heavy. . I have received a premium on a coulter made in this shape.

MILL PICKS

Mill picks are very easily dressed and hardened, the whole trick in this case, as in many others, lies in the right heat of the steel. Be careful not to heat to a higher than a red heat. Dress the pick and temper with a low heat, when the color is dark yellow the temper is right, if the steel is of the right kind. No other hardening compound is necessary than water. After a little experience any smith can do this work first class.

A smith once wanted to buy my receipt for tempering. He believed I had a wonderful prescription, or I could not succeed as I did. I told him I used only water, but he insisted that I was selfish and would not reveal it to him.

If tools and receipts would do the work there would be no need of experienced mechanics. Tools and receipts are both necessary, but it must be a skilled hand to apply them.

HARDEN FILES

The best way to harden files is to have a cast iron bucket filled with lead. Heat it until the lead is red hot, then plunge the file into this, handle up. This will give a uniform heat and the file will not warp so easy if the heat is right. In cooling the file off, use a box four or five feet long with salt water in, run the file back and forth endwise, not sideways, that will warp the file, take it out of the water while yet sizzling. Now, if warped, set it between a device so that you can bend it right. While in this position sprinkle water over where you straighten until cold and the file will be right.

HARDEN TAPS AND DIES

Heat the tap or die to a red cherry, cool off entirely in water, brighten with an emery paper. Now, hold over a hot iron until the tap or die has a dark straw color, then cool off. If a light tap, the temper can be drawn over a gaslight, using a blowpipe.

BUTCHER KNIFE

To make a butcher knife, one smith will simply take an old file, shape it into a knife, and harden. The best way to make a knife is to first draw out a piece of iron ¾ inch wide and $\frac{1}{16}$ of an inch thick, twice the length of the knife. Prepare the steel the same width as the iron, ⅛ of an inch thick, weld this steel in between the iron. This will make a knife that will not break. When ready to harden heat to a low red heat, cool off entirely in water. Brighten and hold over a hot iron until brown, then cool off.

The steel should be good tool steel, a flat file will do, but the cuts must be ground or filed off entirely before you touch it with the hammer, for if the cuts are hammered in they will make cracks in the edge of the knife, and the same will break out.

HOW TO REPAIR CRACKED CIRCULAR SAWS

If a circular saw is cracked it can be repaired so that the crack will go no further, and if the crack is deep, it can be so remedied that there will be no danger in using it. Ascertain the end of the crack, then drill a $\frac{3}{16}$-inch hole so that the crack will end in that hole. Countersink on each side and put in a rivet. Don't let the rivet stick its head over the face of the saw.

If the crack is deep put another rivet about half an inch from the edge. If the saw is too hard to drill, heat two irons about 1¼ square or round, square up the ends and set the saw between the ends so that they will meet over the place where the hole is to be drilled. When the saw is dark blue, the temper is out. It might be a possibility that this will spring the saw in some cases, therefore, I advise you to try drilling the hole without any change in temper. Prepare a drill that is harder than usual, use no oil, but water.

HOW TO PREVENT A CIRCULAR SAW FROM CRACKING

The reason why a circular saw cracks is, in most cases, incorrect filing. In filing a saw, never let a flat file with its square corners touch the bottom of the teeth you are filing; if you do, you will make a short cut that will start the crack. The best way is to gum the saw in a saw gummer or on an emery wheel, or use a round-edged file.

HOW TO SEW A BELT

Belts can be riveted, sewed, or hooked together. A new leather belt should not be riveted, because such a belt will stretch and have to be cut out and sewed over quite often at first. There are hooks made of steel for belt sewing, these are all right when the pulleys are not less than six inches in diameter and the speed is slow. In using these hooks be careful not to bend them too sharp or drive the bends together too hard; in so doing they will cut through the leather and pull out. Lacing is the best for all kinds of belts.

In sewing a belt with lacing, first punch with a punch made for this purpose, holes in proportion to the width. Don't punch them too close to the ends. Begin sewing in the center holes and start so that both ends of the lacing will come out on the outside of the belt. Now sew with one end to each side, and be careful not to cross the lacing on the side next to the pulleys. The lacing should be straight on that side. When the belt is sewed punch a small hole a little up in the belt to receive the last end of the lacing; the last end should come out on the outside of the belt. In this end cut a little notch about three-fourths through the lacing close to the belt, and then cut the lacing off a quarter of an inch outside of this notch. This notch will act as a prong and prevent the lacing from pulling out. Tap it lightly with a hammer above the seam to smooth it down.

POINTS ON BELTS

In placing shafts to be connected by belts, care should be taken to get the right working distance one from the other. For smaller belts 12 to 15 feet is about the right distance. For large belts, a greater distance is wanted. The reason for this is that when pulleys are too close together there is no sag in the belts and they must therefore be very tight in order to work.

Belts should not have too much sag, or they will, if the distance between the pulleys is too far apart, produce a great sag and a jerking motion which will be hard on the bearings.   Never place one shaft directly over another, for then the belts must be very tight to do the work, and a tight belt will wear out quicker and break oftener in the lacing than a loose one; besides this the bearings will give out sooner.

If a belt slips use belt oil or resin, or both.

BOB SHOES

In repairing old bob sleds is is difficult to find shoes to suit.   But in every case the shoe can be fitted to suit without touching the runner.   The trick here as in many other cases in the blacksmith business, lies in the heating.   Any shoe can be straightened or bent to fit the runner if only heated right.   A low cherry-red heat and a piece if iron to reach from the crooked end of the shoe and far enough back to leave a space between where it wants to be straightened.   Now put it in the vise and turn the screws slowly and the shoe will stand a great deal.   If too straight, put the shoe in between a couple of beams so that you can bend it back to the right shape.   Remember the heat.

I have put on hundreds and never knew of a shoe that broke when the heat was right.   I must confess,

however, that my two first shoes broke, but I think I
learned it cheap when I consider my success after
that. The shoe should fit the runner snug. Ironing
bobs is a very simple and easy thing, every blacksmith, and even farmers sometimes, are able to iron
their own sleds fairly well, and I don't think it will be
of much interest for the readers of this book to treat
that subject any further.

AXES AND HATCHETS

Dressing axes is quite a trick and few blacksmiths
have mastered it. It is comparatively easy when one
knows how. I have several times already warned
against over heating and if this has been necessary
before, it is more so now in this case. In heating an
ax do not let the edge rest in the center of the fire, it
will then be too hot at the edge before it is hot enough
to hammer it out. Place the edge far enough in to let
it over the hottest place in the fire. Go slow. When
hot, draw it to the shape of a new axe, don't hammer
on one side only. In so doing the ax will be flat on
one side and curved up on the other. If uneven trim
it off; trim the sides also if too wide; don't heat it
over the eye; be sure you have it straight. When
ready to harden, heat to a low red heat and harden in
luke warm water. The heat should be only brown if
it is a bright sunny day. Brighten and look for the
temper. You will notice that the temper runs uneven;

it goes out to the corners first, therefore dip them (the corners) deeper when cooling, and with a wet rag touch the place on the edge where the temper wants to run out. Some smiths, when hardening, will smear the ax with tallow instead of brightening it, and hold it over the fire until the tallow catches fire, then cool it off. This is guess work, and the axe is soft in one place and too hard in another. The best way is to brighten the ax and you can see the temper, then there is no guess work about it. When blue cool it partly off and then while the ax is still wet you will observe under the water or through the water a copper color. This color will turn blue as soon as the ax is dry, and is the right color and temper. Cool it slowly, don't cool it off at once, but let it cool gradually, and it will be both hard and tough.

By this simple method I have been very successful, breaking only three per cent, while no new ax of any make will ever do better than ten per cent. Some will even break at the rate of twelve and thirteen per cent.

The ax factories, with all their skill and hardening compounds, have to do better yet to compete with me and my simple method.

WELL DRILLS

Well drills are made of different sizes and kinds. Club bits and Z bits. How to dress: heat to a low red heat. If nicked or broken, cut out, otherwise draw it out to the size wanted. The caliper should touch the lips of the bit when measured diagonally so that the bit has the size on all corners. Heat to a low red heat and harden, the temper to be from dark straw color to blue according to the kind of drilling to be done. The trick, in two words, low heat.

GRANITE TOOLS

By granite tools is meant tools or chisels used by granite or marble workers for cutting inscriptions on tombstones.

When a man understands how these tools are used it is easier to prepare them. These are the kind of tools where an unusual hardness is required. The hammer used in cutting with this chisel is very small, and the blow would not hurt your nose, so light it is, therefore they will stand a high heat and temper. The chisels should be very thin for this work. When dressed and ready to harden, heat to a red heat and harden in the following solution: one gallon soft water, four ounces salt. Draw the temper to a straw color.

A blacksmith once paid a high price for a receipt for hardening granite tools. The receipt was, aqua, one gallon; chloride of sodium, four ounces. This receipt he kept as a secret and the prepared compound he bought at the drug store, thus paying 50 cents for one gallon of water and four ounces of salt. The real worth is less than a cent. It is said he succeeded remarkably well with his great compound, which he kept in a jug and only used when anything like granite tools were to be hardened. The reason why he succeeded so well was because of his ignorance concerning his compound, not because it was not good enough. I hold that it is one of the best compounds, in fact, the best he could get. People in general like to be humbugged. If they only get something new or something they don't know anything about, then they think it wonderful.

Salt and water should be called salt and water, and be just as much valued. Let us "call a spade a spade," the spade will not be more useful by another name, nor will it be less useful by calling it by its proper name.

CHAPTER IV

WHEN vehicles were first used is hard to tell, but we know that they have been used for thousands of years before the Christian era. It is easy to imagine how they looked at that time, when we know how half-civilized people now make wagons. The first vehicle was only a two-wheeled cart called chariot. Such chariots were used in war and that it was a case of "great cry and little wool" is certain.

The blacksmith used to be the wagon and carriage maker. Now it is only a rare case when a blacksmith makes a carriage, and when it happens most of the parts are bought. In 1565 the first coach was made in England.

Now there are hundreds of factories making wagons and carriages and parts of them for repair use by blacksmiths and wagon makers. It is no use for any blacksmith or wagon maker to compete with these factories. We have neither the means nor the facilities to do it, and have to be content with the repairs they need. The most important repairs are the setting of tire, welding and setting axle stubs,

SETTING TIRE

Wagon tire is often set so that more harm than good is done to the wheel.

In setting tire the first thing to do is to mark the tire. Many blacksmiths set tires without marking the tire. This is poor work. In order to do a good job the tire should be set so that it is in the same place it had. There are generally some uneven places in the fellows and when the tire is set the first time, it is hot all around and will settle down in these low places. Now, if the tire is not marked and set back in its exact bed, it will soon work loose again, and it is liable to dish the wheel too much as it don't sink into its place, but is held up in some places. Another thing, when a tire is worn so that it becomes thin it will settle down on the outside, especially when the wheel is much dished. Now if you reverse the tire it will only touch the fellow on the inner edge of the wheel, and leave an open space between the fellow and the tire on the outside. When a wheel has bolts every smith knows that it will make trouble for him if he don't get the tire back where it was. In every case take a file or a chisel and cut a mark in the tire near to the fellow plates, cut also a light mark in the fellow. These marks are to be on the inside of the wheel: 1, because it will not be seen on that side; 2, because in putting the tire on, the wheel should be placed with that side up. If there are nails in the tire cut them off with a thin chisel so that it will not mark the fellow, or drive them into the fellow with a punch. Next, measure the wheel with the gauge (the wheel is supposed to be right, not fellow bound nor any spokes loose in the tenon). This done, heat the tire and shrink it. If the wheel is straight give it half an inch draw, sometimes even five-eighths if the wheel is heavy and strong. But if the wheel is poor and dished, do not give it more than one-fourth-inch draw. One tire only with a little draw can be heated in the forge, but if there is more than one tire heat them outside in a fire made for this purpose, or in a tire heater.

There are different ways of cooling the tire. Some smiths have a table in a tank, they place the wheel on the table and with a lever sink both wheel and tire in the water. There are many objections to this. 1, You will have to soak the whole wheel; 2, it is inconvenient to put the tire on; 3, in order to set the tire right, it is necessary to reach the tire from both sides with the hammer; 4, when spokes have a tendency to creep out, or when the wheel is much dished, the wheel should be tapped with the hammer over the spokes. Now, to be able to perform all these moves, one must have, first, a table; this table to be about twelve inches high and wide enough to take any wheel, with a hole in the center of table to receive the hub. On one side you may make a hook that will fall over the wheel and hold the tire down while you get it on. Close to this table have a box 5½ feet long, 12 inches wide and 12 inches deep. On each side bolt a piece of two by six about three feet long. In these planks cut notches in which you place an iron rod, run through the hub. On this rod the wheel will hang. The notches can be made so that any sized wheel will just hang down enough to cover the tire in the water. In this concern you can give the wheel a whirl and it will turn so swift that there will be water all around the tire. It can be stopped at any time and the tire set right, or the spokes tapped. With these accommodations and four helpers I have set six hundred hay rake wheels in nine and one-half hours. This was in a factory where all the tires were welded and the wheels ready so that it was nothing but to heat the tires and put them on. I had three fires with twelve tires in each fire. An artesian well running through the water box kept the water cool.

If the fire is not hot enough to make it expand a tire puller is needed. A tire puller can be made in many ways and of either wood or iron. Buggy tire is more particular than wagon tire and there are thousand of buggy wheels spoiled every year by poor or careless blacksmiths. In a buggy tire one-eighth of an inch draw is the most that it will stand, while most wheels will stand only one-sixteenth. If the wheel is badly dished don't give it any draw at all, the tire should then measure the same as the wheel, the heat in the tire is enough.

If the wheel is fellow-bound cut the fellows to let them down on the spokes.

If the spokes are loose on the tenon wedge them up tight.

BACK DISHED WHEEL

For a back dished wheel a screw should be used to set the wheel right. Place the wheel on the table front side up. Put wood blocks under the fellow to raise the wheel up from the table. Place a two by four over the hole under the table; have a bolt long enough to reach through the two by four and up through the hub, a piece of wood over the hub for the bolt to go through; screw it down with a tail nut. When the wheel is right, put the tire on. The tire for such a wheel should have more draw than for a wheel that is right.

If a buggy wheel has been dished it can be helped a little without taking the tire off. Place the wheel on the anvil so that the tire will rest against the anvil. Don't let the tire rest lengthwise on the anvil. If you do, the tire will be bent out of shape when you begin to hammer on it. Use the least surface possible of the anvil and hammer on the edge of the tire; the stroke of the hammer to be such that the blow will draw the tire out from the fellow. A tire too tight can be remedied this way.

When bolting a wheel the tire will be out of place unless the tire has been shrunk alike on both sides of the fellow plates. A smith used to setting tires will be able to get the holes almost to a perfect fit. If a tire is too short, don't stretch it with a sharp fuller that will cut down into the tire, when the tire is a little worn it will break in this cut. Draw it out with a wide fuller and smooth it down with the hammer. If it is

much too short, weld in a piece. This is easily done.
Take a piece of iron ¼-inch thick, the width of the tire
and the length needed, say about three inches. Taper
the ends and heat it to a red heat. Place it on the tire
in the fire and weld. This will give material for
stretching.

If the wheel has a strong back dish it cannot be set
right to stay with the tire alone, as a bump against the
fellow is apt to throw the dish back. It is therefore
safer in all back dished wheels to take the spokes out
of the hole and set them right by wedges in the end of
the spokes. These wedges should not be driven from
outside in but be placed in the end of the spoke so that
they will wedge into the spoke when the same is
driven back into its place. Use glue.

HOW TO PUT ON NEW TIRE

When you have the bar of either steel or iron for the
tire, first see if it is straight, if not be sure to make it.
Next place the tire on the floor and place the wheel on
top of the tire, begin in such a way that the end of the
fellow will be even with the end of the tire. Now roll
the wheel over the tire. If a heavy tire cut it three
inches longer than the wheel, if a thin tire, two inches.
Now bend the tire in the bender. Measure the wheel
with the gauge, then measure the tire; if it is a heavy
wagon tire and a straight wheel cut the tire one-fourth
of an inch shorter than the wheel. If it is a buggy
tire cut it the size of the wheel. In welding these
tires they will shorten enough to be the size wanted.

HOW TO WELD TIRES

There are many different ideas practiced in welding tires. One smith will narrow both ends before welding; another will cut the edges off after it is welded. This is done to prevent it from spreading or getting too wide over the weld. I hold that both these ideas are wrong. The first one is wrong because when the ends are narrowed down it is impossible to make them stay together until the weld is taken, especially if it is a narrow tire. The second idea is wrong because it cuts off the best part of the weld and weakens it. Some smiths will split the tires, others will rivet them together. This is done to hold the tire in place until it has been welded. There is no need of this trouble, but for a new beginner a rivet is all right.

I shall now give my experience in welding tire, and as this experience has been in a factory where thousands of wheels are made yearly, I suppose it will be worth something to the reader.

When the tire is ready to weld draw down the ends and let them swell as much as they want to. Now let the helper take the end that is to lay on top and pull it towards the floor, the other end to rest on the anvil. This will give that end a tendency to press itself steadily against the lower end. Next place this end on top of the other end. The ends must now be hot enough to allow them to be shaped. You will now notice that the top end is wider than the tire, so is the lower end. The tire is to be so placed that the swelled parts reach over and inside of each other a little. Now give a couple of blows right over the end of the under tire. Next tap the swelled sides down over the tire. This will hold the tire together so that it cannot slip to either side, and the swelled end of the under tire will prevent it from pulling out. If the top end has been so bent that it has a tendency to press down and out a little, the tire will now be in a good shape to weld.

Before you put the tire into the fire, let me remind you of what I have said before about the fire. Many blacksmiths are never able to weld a tire tight on the outside because of a poor, low, and unclean fire. If the fire is too old or too fresh it will not give a good heat for welding tire. If you have a good big fire high up from the tuyer, then you are all right. Place your tire in the fire and proceed as follows: No matter whether it is an iron or soft steel tire, sand is the best welding compound and nothing else should be used; but if you lose the first heat then borax might be used as it will prevent the tire from scaling and burning. When you have the right heat, place the tire on the anvil this way; let the tire rest against the inside edge of the anvil. If the lower end of the tire is allowed to come down on the anvil it will cool off and can never be welded that way. Now hold the tire this way until you have the hammer ready to give the first blow. Then let the tire down and strike the first blows directly on top and over the end of the under end. This is important and if the first blows are not directed to this very place the lower end will be too cool to weld when you get to it. Next weld down the upper end, this done turn the tire on edge and while it is in a welding heat come down on it heavy with hammer, if a buggy tire, and with a sledge and hammer if a heavy wagon tire. Hammer it down until it is considerably narrower over the weld as it will swell out when dressed down. This way the weld has all the material in the iron and the lapped lips will help hold the weld together. A very poor smith can weld tires to stay in this manner. The edges should be rounded off with the hammer and filed to make the tire look the same over the weld as in the iron. If there should be any trouble to weld a steel tire place a little steel borings over the weld and use borax.

A blacksmith in Silver Lake, Minn., working for a wagon maker of that place, when welding a tire failed entirely after half a dozen attempts, and he got so angry that he threw the tire down on the floor with all his might. It happened to crush the wagon makers big toe. This was more than the otherwise good-natured man could stand, and instantly the smith was seen hurled through an open window—the wagon naker attached. Result: separation and law suit. ll this because the smith had not read my book.

When a light buggy tire is to be set mistakes are ften made in measuring the tire. The tire is too ght in itself to resist the pressure of the gauge. The nith tries to go it light and if there is not the same essure in measuring the tire there was in measuring e wheel, it will not give the same results; and when e tire is put on it is either too tight or too loose. worked for many years on a tool to hold the tire steady in order to overcome this trouble. The only device that I have ever seen for this purpose before is the anvil close up to the forge, one side of the tire on the forge, the other on the anvil. This arrangement would crowd the smith, roast his back and expose him to ridicule, but it will not help to ruin the tire.

The tool I invented is a tire holder made of cast iron. It consists of a standard or frame with a shank in to fit in the square hole in the anvil; in the standard is a slot hole from the bottom up. On the back of the standard are cogs on both sides of the slot hole. Through this goes a clutch hub with cogs in to correspond with the cogs in the standard. On the outside of the standard is an eccentric lever. Through this lever is a tapered hole to fit over the clutch hub. This lever is tapered so that it will fit different thicknesses, while the cogs and eccentric lever will adjust it to different widths. This device is so cheap that any smith can afford to have it.

Next time you buy a quart of whisky sit down and figure out which will do you more good, my tire holder or the whisky. Figure 7 is an illustration of my holder. This tool is better than an advertisement in your local paper, of which the following story will convince you. A blacksmith in a prohibition county in a northern Iowa town got into the habit of going over to a Minnesota town for a keg of beer every month. On one of his periodical visits to this place he saw a crowd of men standing around a road grader in the road. As he approached he found that the grader had a serious break-down and the men were just discussing the possibilities of getting the grader repaired in the village shops. One said no smith could do it, another thought they could if they only had tools. "I know a man," said one in the party,

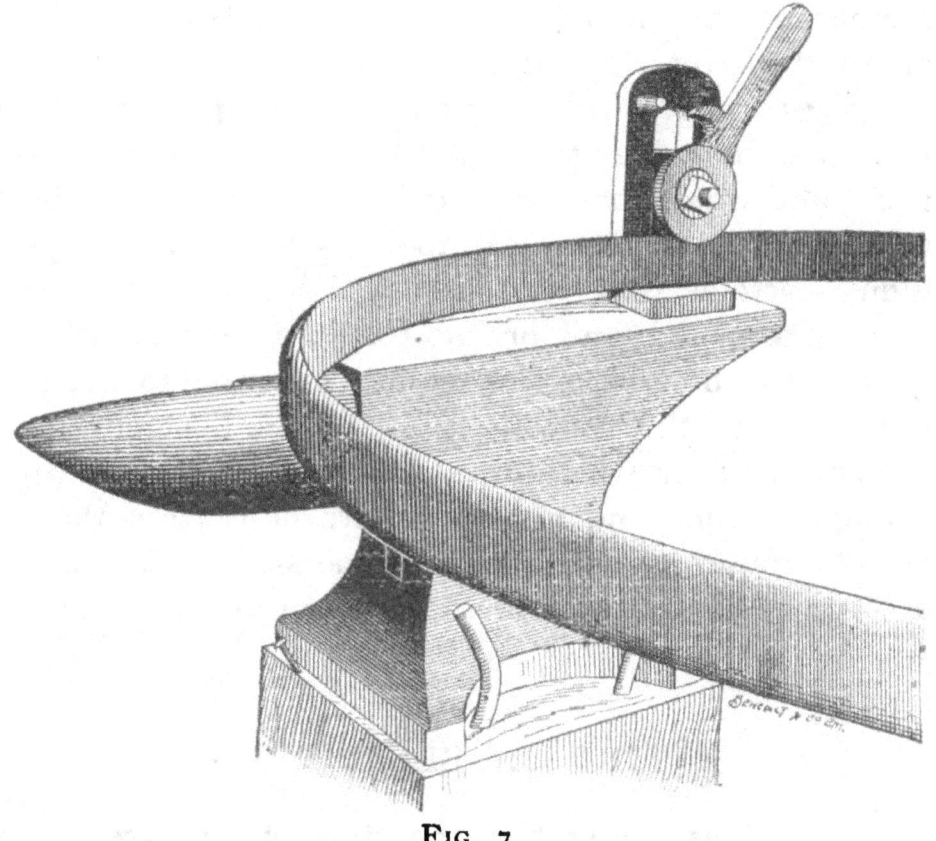

FIG. 7

HOLMSTROM TIRE HOLDER

"that can if any man can, and he has tools I am sure. I was over to his shop the other day to have my buggy tire set, and mind you, he had the slickest tool you ever saw to hold the tire in; I never saw a tool like that before." "Well," said one, "that has nothing to do with this case." "Yes it has," said the road boss, "my father always used to say, 'A mechanic is known by the tools he uses,' and when a smith has good tools in one line, he has them in another, and I shall give this man a chance."

Our traveling smith had heard enough. This was a temperance and tool lecture to him, he began to think of all the trips he had made to this town. Twelve trips a year, three dollars a trip for liquor and the time lost must be worth two dollars per day. He figured it out and would have turned back if he had not been so close to the place. He took a glass of beer but it didn't taste as usual and he asked for a cigar. With this he returned, and on the road home swore off for good. He bought a tire holder at once to start in with, and by this time he is one of the best smiths in the country, always at his stand ready to do the work brought to him, and his customers now know that he is to be found in his place, with tools of all kinds and a sober hand to use them with. Do thou likewise.

TIRE IN SECTIONS

Many of us remember the time when tires were made in sections and nailed on, at this time the wheels were more substantially made, because the tire could not be set as tight as it is now, and the wheel had to be made so that it would stand the usage almost independent of the tire. Our endless tire is a great improvement over the tires made in sections. The wagon tires as they are made now are, I think, as near right as they can be, in regard to size of iron, in proportion to the wheel. But it is different with buggy tires. I hold that they are all made too light to be of any protection to the fellows. I understand the reason why they are made this way, but if a man wants a light rig, let that be the exception and not the rule.

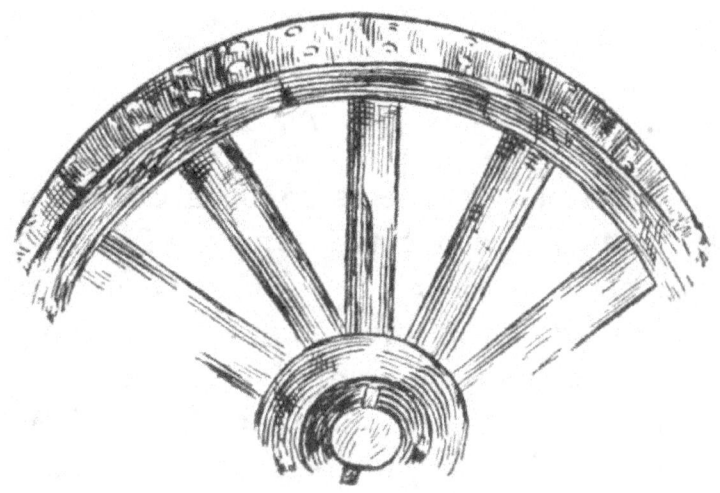

Tire should not be less than one-fourth of an inch thick for seven-eighths wide, and five-sixteenths for an inch wide and over.

EXPANSION OF THE TIRE

A tire four feet in diameter will expand two inches and a quarter, or three-sixteenths of an inch to the foot. Steel tire expands less. This is the expansion of red heat. If heated less it expands less, but it is no trouble to make the tire expand for all the draw it needs.

A furnace for tire heating comes handy in cities where there is no chance for making a fire outside, but every smith that has room for a fire outside will do better to heat the tire that way. Don't build a tire-heating furnace in the shop if wood is to be used for fuel, because the heat and smoke will turn in your face as soon as the doors of the furnace are opened.

WELDING AXLES

When a worn buggy axle is to be stubbed, proceed as follows: First, measure the length of the old axle. For this purpose take a quarter inch rod of iron, bend a square bend about an inch long on one end. With this rod measure from the end of the bearing, that is, let the hook of your rod catch against the shoulder at the end where the thread begins, not against the collars, for they are worn, nor should you measure from the end of the axle, for the threaded part is not of the same length. Now place your stub on the end of the axle and mark it where you want to cut it off. Cut the axle one-fourth inch longer than it should be when finished. Next heat the ends to be welded and upset them so that they are considerably thicker over the weld; lap the ends like No. 1, Figure 6, weld and use sand, but if the ends should not be welded very well then use borax. These stubs are made of soft steel, and will stand a higher heat than tool steel, but remember it is steel. If the ends have been upset enough they will have stock enough to draw down on, and be of the right length. If this is rightly done one cannot tell where the weld is. Set the axle by the gauge, if you have one, if not, by the wheels.

AXLE GAUGE

A gauge to set axles by can be made in this way: When you have set an axle by the wheels so that it is right, take a piece of iron 1¼ x ¼, six feet long, bend a foot on this about six inches long, with a leg on the other end. See No. 5, Figure 8; the leg to be movable and set either with a wedge or a set screw to fit for wide and narrow track. The gauge to be set against the bottom side of the axle. The pitch to be given a set of buggy wheels should be from one to one and one-half inches. I would recommend one and a half inches. This will be enough to insure a plumb spoke when the vehicle is loaded. It will also insure safety to the rider from mud slinging. By pitch, I mean that the wheels are one and a half inches wider at the upper rim than they are down at the ground. Every smith ought to have a gauge of this kind, it is easy to make and it saves a lot of work, as there is no use of the wheels being put on and an endless measuring in order to get the axle set right.

GATHER GAUGE

By gather I mean that the wheels should be from one-fourth to one-half an inch wider back than in front. Don't misunderstand me now. I don't mean that the hind wheels should be wider than the front wheels, I mean that a wheel should have a little gather in front, as they are inclined to spread and throw the

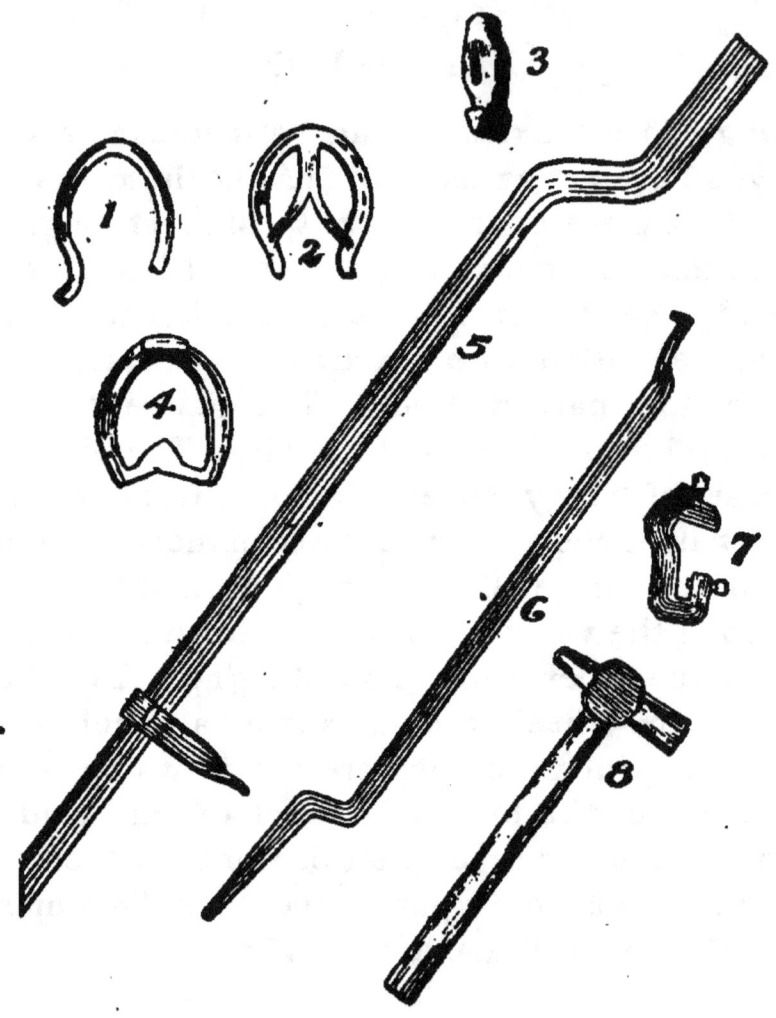

FIG. 8

bearing on the nut, while, if they have a little gather, they will run right, and have a tendency to throw the bearing on the collars of the axle. If they do they will run more steady, especially when the axle is a little worn.

A gauge for this purpose can be made like Figure 8, No. 6. This gauge to be fitted to the front side of the axle when you make it. It can be made of 1 x ¼ about three feet long, the forked end to reach the center of the axle. With these two gauges axles can be set right without the wheels.

"The sluggard will not plow by reason of cold; therefore he shall beg in harvest and have nothing."—Proverbs.

CHAPTER V

HOW TO MAKE PLOWSHARES

HERE are two kinds of shares: lip shares and bar shares, and they must be treated differently. We will first treat of bar shares. The first thing to do when a plow is brought for a new lay is to look over the condition of the landside. By landside is meant the bar to which the share is welded. Now if this bar is worn down so that you think it too weak to stand for a new share, then make a new one.

HOW TO MAKE A LANDSIDE

For a 14-inch plow take 2½ x ⅜, or 2½ x $\frac{7}{16}$. For a 16-inch plow, use 2½ x $\frac{7}{16}$, or 3 x $\frac{7}{16}$ common iron. Cut the iron diagonally at the point. This will prepare a point on each side of the cut; that is, you had better cut out two landsides at a time. But if you do not want to do that, then cut the iron off square. Next take a piece of common iron 3 x ¼, 13 inches long for a shin; cut this diagonally, and it will make shins for two. Some plow factories use steel for shins, but that

88

is not necessary, for it will not make the plowshare
any better, but, on the other hand, will be quite a

JOHN DEERE, THE INVENTOR OF STEEL PLOWS

bother when you want to drill a hole for a fincoulter if
it is hardened.   Place this shin on the land side of the
landside, and weld.   In preparing the shoulder of the

shin for the plate use a ship upsetter. See No. 3, Figure 8.

Not one out of 500 blacksmiths have this tool. Every smith should have one. You cannot do a good and quick job without it.

When you shape the point of the landside hold it vertical, that is, the edge straight up and down, or plumb. If you don't do this, there will be trouble in welding, especially if you have held it under. Then it will lean under the square when welded, and in such a case it is hard to get a good weld, and if you do you will break it up when you attempt to set it to the square. Another thing, don't make much slant on the landside up at the joint, for, if you do, you can never weld the share good up there. Give more slant towards the point. Be sure to have the right curve. It is very important to have the landside right: 1, Because it is the foundation for the plow; 2, if the landside is right the start is right, and then there is no trouble to get the share right. When finished place the old landside on top of the new, with the upper edges even; don't go by the bottom edges, as they are worn. Now mark the hole. You may leave the front hole for the foot of the beam this time. When holes are drilled, then put a bolt through the hole of the foot of beam and landside; now place the plow on the landside and measure 14 inches from the floor up to the beam. In this position mark the front hole of the foot of the beam. If the beam has been sprung up you will now have remedied that. So much about a new landside. On the other side, if the old landside is not

too much worn to be used, then repair as follows:
Take a piece of ⅜-inch thick flat iron the width of the
landside about ten inches long.　Cut one end off diagonally, this end to be flattened down.　Why should this
end be cut diagonally?　This piece of iron is to be
placed on the inner side of the landside and as far back
as to cover the hole that holds the plate.　Now, if this
iron is cut square off, and left a little too thick on that
end, it will cut into the landside and weaken it; but if
cut diagonally and drawn out thin it will not weaken,
nor can it break when cut in this manner.　To be sure
of a good strong weld, upset over the weld.　I hold
that this is the most important thing in making a new
lay.　"No hoof, no horse" — no landside, no plow.
There are only a few blacksmiths recognizing this fact.
Most of the smiths will simply take a piece of iron
about half an inch square and weld it on top of the
point.　This is the quickest way, but it is also the
poorest way, but they cannot very well do it in any
other way, for if you have no shin upsetter to dress
and shape the shoulder for the plate, then it is quite a
job to repair any other way.　There are three reasons
why a landside cannot be repaired with a patch on top
of the point: 1, The shin or shoulder in an old landside
is worn down sometimes to almost nothing, and the
only way to get stock enough to make a good shoulder
is to put a good-sized piece of iron on the inside, back
and behind this shoulder.　If a new plate is to be put
on and this is not done, you will have to draw down
the plate to the thickness of the old shoulder, and in
such a case the plate will add no strength to the share.

2, The landside is, in many cases, worn down on the bottom to a thin, sharp edge, and by placing the piece on top the landside will be as it was on the bottom side, where it ought to be as thick as you can make it. 3, The weakest place in the landside is just at the shoulder of the shin, and by placing the piece on top it will not reach over this weak place, and with a new long point on, the strain will be heavier than before, and the landside will either bend or break. I have in my experience had thousands of plows that have been broken or bent on account of a poorly-repaired landside. Blacksmiths, with only a few exceptions, are all making this mistake.

The landside is to the plow what the foundation is to the house. No architect will ever think of building a substantial house without a solid foundation. No practical plowsmith will ever try to make a good plow without a solid landside.

For prairie or brush breakers, where no plate is used, it will be all right to repair the landside by placing a piece of iron on top of it, provided it is not much worn, and the patch reaches back far enough to strengthen the landside. But even in such cases it is better to lay it on the inner side.

LANDSIDE POINT FOR SLIPSHARE

We have now learned how to prepare the landside for a solid or long bar share. We shall now learn how to make a landside point for slipshares. There are smiths that will take the old worn-out stub of a slip-share point, weld a piece to it, and then weld the share on. This is very ridiculous and silly. There is nothing left in such a point to be of any use. Make a new one; be sure to make it high enough—at least half an inch higher than the share is to be when finished. This will give you material to weld down on. If the landside is not high enough the share will be lower—that is, the joint of the lay will be lower than the joint of the mouldboard, and it should be the other way.

PLOW OF 200 YEARS AGO

On this point many an old smith and every beginner makes mistakes, and not only in this case, but in everything else. Whatever you have to make, be sure to have stock enough to work down on, and you will be all right. It is better to have too much than not enough.

In shaping the point remember to hold it perpendicular, and give very little slant up at the joint, but more towards the point. If too much slant up at the joint there will be difficulty in welding it. Remember this. Don't make the point straight like a wedge; if you do the share will be above the frog. Give it the same circle it had, and the share will rest solid on the frog. This is another important point to remember:

The lay will not have the full strength if it don't rest on the frog, and it will not be steady, and the plow will not run good, for in a few days the share flops up and down.

When a 14-inch share is finished the point, from the joint of the share to the extreme end of the point, should be 11 inches, not longer, and for a 16-inch lay, 12 inches, not longer. The point acts as a lever on the plow, and if it is too long the plow will not work good, and it is liable to break. Shape the point so that when you hold it up against the plow it will be in line with the bottom of the landside, but about half an inch wider than the landside to weld on. If it is a plow where the point of the mouldboard rests on the landside point, and it is a double shin, then cut out in the landside point for the point of the mouldboard to rest in. See No. 1, Figure 9. This will be a guide for you when welding the share, and it will slip onto the plow easier when you come to fit it to the same. I think enough has been said about the landside to give the beginner a good idea of how to make one. And if the landside is right, it comes easier to do the rest. In making a plowshare there are many things to remember, and one must be on the alert right along, for it will give lots of trouble if any point is overlooked.

We will now weld a share to a long bar landside. The landside having been finished and bolted to the beam or its foot, or to a standard, the share is to be shaped to fit. Hold the share up to the plow. First look if the angle for the point is right in the share; if not, heat the share, and if under the angle wanted

upset up at the joint; if over the angle wanted, drive
it back at the point.   In doing this hold the edge of
the share over a wooden block instead of the anvil, so
as not to batter the thin edge of the share.   If the
share has been upset so that it has a narrow rib along
the point where it is to be welded, draw this down and
make it level.   In most blank shares the point should

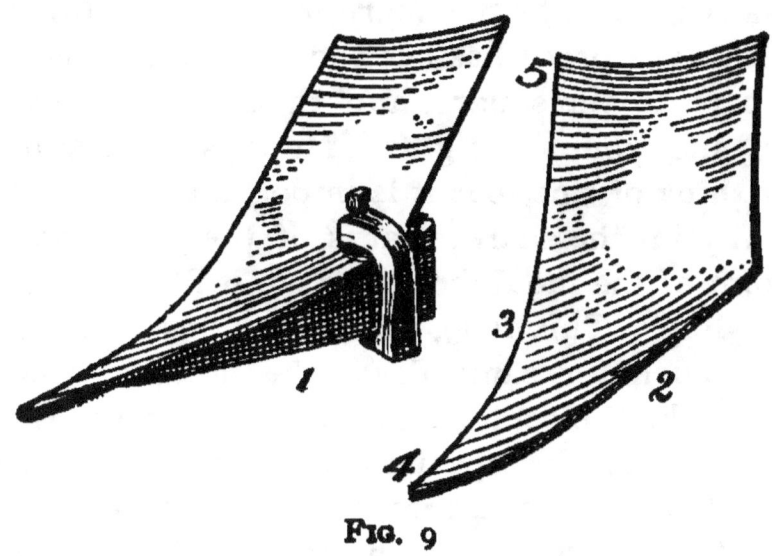

Fig. 9

be raised to fit the landside point, so that when the
same is placed on the floor the edge of the share will
follow the floor or leveling block (if you have it), from
the heel right up to the point, then it will be easy to
make the edge come down to the square in finishing it
up.   If this is not done the edge of the share from the
throat back will generally be too high.

In Figure 9 two shares are represented, one with the
landside point on ready for welding.   In this share the
point of the same has been raised so that the share

comes down to the square in the throat.   The other is a blank share, straight in the point between Nos. 4 and 5, resting on the extreme heel and point with gap between the edge of share and floor at No. 3.   In most blank shares the point is too straight, and the point too much bent down at No. 4.   Bend the share so that the whole length from heel to point will follow the floor.   When the share is held in a position as shown in this cut, don't fit the share to the brace, for in most old plows the brace has been bent out of shape.   Fit the share to the square, and then fit the brace to the share, and you are right.   Many a blacksmith will never think of this, but it is important.

Next joint the share; that is, if the joint does not fit the joint of the mouldboard, make it fit either by filing or grinding.   This done, make the holes, and when you center-punch for same draw the holes a trifle; that means make the center mark a little towards the inner side of the mark, especially for the hole next to the point.   This is also an important point overlooked by most blacksmiths.   The holes that hold the joints together should act as a wedge.   If they don't the joints will pull apart and leave a gap between, where dirt and straw will gather, and if a slipshare the share will soon work loose and the plow will flop.

The holes having been punched and countersunk, the share should be bolted to the brace.   Next put on the clamp.   It is not necessary that the clamp should be put on while the share is on the plow.   I never do that.   I used to for many years, but there is no need of doing it, for if the share has the right angle it must come to its place when even with the point on the outside, and a cut should be made in the landside just at the place where the point of the mouldboard rests on same, this cut will also be a guide.

Now a few words concerning the clamp. Figure 8, No. 7 illustrates a clamp for this purpose. The set screw at the bottom serves to hold the landside from leaning over or under, while the setscrew at the upper end holds the share against the point. If this clamp is rightly made it works splendid. The clamp should be placed over the plowshare up at the joint, because the first heat or weld should be on the point. Some smiths —well, for a fact, most smiths—take the first weld up at the joint. This is wrong. The point should be welded first. Then you have a chance to set the share right and fit it snug to the point the whole way up. You cannot make a good weld if the share does not fit snug against the landside point, to prevent air and cinders from playing between. Further, the share should be upset over the weld, when this is not done in the blank share; the lower corner of the share will protrude over the landside. This should be dressed down smooth. The next weld should be taken up at the joint. For welding compound use steel borings and scales from either steel or iron.

After you have moistened the place where the weld is to be taken with borax, then fill in between the share and point with steel borings, and on top of this a little steel or iron scales. Do not buy any welding compound of any kind, because if you learn to know what you have in the shop you will find that there never was a welding compound made to excel borax, steel scales, steel or iron borings, and powdered glass. All these you have without buying.

In heating go slow. If you put on too strong blast the share will burn before the iron is hot enough to weld. When ready to weld let your helper take with a pair of tongs over the share and landside to hold them tight together while you strike the first blow. Use a large hammer and strike with a pressure on the hammer the first blows, until you are sure it sticks; then come down on it with force.

I have made it a practice, no matter how good this weld seems to be, to always take a second weld. This weld to be a light one. The share and landside are after the first weld settled, so it takes very little to weld them then. On the other hand, the first weld might look to all appearances solid, but it is not always. With this precaution I never had a share that ripped open in the weld, while it is a rare thing to find a share made by a blacksmith that does not rip. Now, then, weld down toward the point. The point should not be allowed to have any twist, for if it does, it will turn the plow over on the side. Now set the edge right, beginning at the heel. If the share is made for hard fall plowing give more suction than for a share for soft spring plowing. Grind and polish before you harden, and after it is hardened touch it up lightly with the polish wheel. Much polishing or grinding after hardening will wear off the case hardening.

SLIPSHARE

We shall now weld a slipshare. When the point is finished hold it to the plow with a pair of tongs while you fit the share. When the share is fitted take the point off from the plow and fasten it to the share with the clamp. As I have said before, there is no need of fastening the share to the landside point with the plow as a guide. If the landside and share are right there cannot be any mistake, and it comes easier to screw them together over the anvil. Now proceed as with a long bar share, and when the weld up at the joint has been taken, fit the share to the plow while hot. Some smiths in preparing the landside point for a slipshare will place the share so that the point is a little too short back where it rests against the end of the plate. This is a bad idea. It is claimed that, in welding, the landside point will swell enough to make it reach up against the plate. This is true, if the landside point is only high enough; but if it is low and you lose a heat in welding, as most smiths do, then your landside point will be both too low and too short. Thousands of shares are made every year that have this fault. Therefore, whatever you are doing have stock enough. It is easy to cut off from the landside while yet hot, but it is difficult to repair if too short. No share will work steadily if the point does not rest right against the plate.

In blacksmithing, every beginner, and many an old smith, makes the mistake of providing less stock than is needed for the work to be done. It is essential to have material to dress down on; and if a heat is lost, or a weld, it will make the stock in the article weaker, and to meet these exigencies there must be material from the start, enough for all purposes.   There is also

a wide difference of opinion as to whether the share should be welded at the point or at the joint first. While I was yet a young man and employed in a plow factory, I had an opportunity to see the different ideas set to a test.   In the factory the practice was to weld the point first. A plowman from another State was engaged, and he claimed that it would be better to weld the share first up at the joint. He was given a chance to prove his assertion, and the result was that 3 per cent of his shares broke over the inner side of the landside at the joint in the hardening, and 10 per cent ripped up in the weld at the same place. These are results that will always follow this method.

The first, because the share was not upset over the weld; the second, because a good weld cannot be taken unless the share is dressed down snug against the point when hot. As far as the number of shares welded per day was concerned, this man was not in it. Still, this man was a good plowman, and was doing better than I ever saw a man with this idea do before. For it is a fact, that out of one thousand plowshares welded by country blacksmiths, nine hundred and ninety will rip up. I have been in different States, and seen more than many have of this kind of work, but, to tell the truth, there is no profession or trade where there is so much poor work done as in black-smithing, and especially in plow work. Blacksmiths often come to me, even from other States, to learn my ideas of making plowshares. On inquiring, I gener-ally find that they weld a piece on the top of the old landside and proceed to weld without touching the share or trying to fit it at all. We need not be sur-prised at this ignorance, when we know that it is only fifty years since John Deere reformed the plow industry entirely and made the modern plow now in use. It is impossible for blacksmiths in the country to have learned this part of their business, in so short a time, successfully. Still, I have seen blacksmiths prosper and have quite a reputation as plowmen, while, for a fact, they never made a plowshare that was, from the standpoint of a practical plowman, right.

CHAPTER VI

HOW TO HARDEN A PLOWSHARE

If the share is of soft center steel, harden as follows: First, heat the whole point to a very low red heat; then turn the share face down, with the heel over the fire, and the point in such a position that it is about two inches higher than the heel. This will draw the fire from the heel along towards the point, and the whole length of the share will be heated almost in one heat. Be sure to get an even heat, for it will warp or crack if the heat is uneven. When the share has a moderate red heat take it out, and you will notice that it is sprung up along the edge. This is the rule, but there are exceptions, and the share is then sprung down. In either case set it right; if sprung up set it down a little under the square; if sprung down set it a little over the square. You cannot with any success set it by a table or leveling block, because this will, first, cool off the edge, second, it must be either over or under the square a little. Therefore, you must use your eye and set the share with the hammer over the anvil. This done, hold the share over the fire until it has a low red heat, as stated before; then plunge it into a tub of hardening compound, such as is sold by the traveling man, or sprinkle the share with prussiate of potash and plunge it into a barrel of salt water.

You will notice that the share will warp or spring out of shape more in the heating than it does in the cooling, if the heat is right. Some smiths never look at the share when hot for hardening, but simply plunge it into the tub, and then they say it warped in hardening, while it was in the heating. If the share is too hot it will warp in cooling also.

HOW TO POINT A SHARE

Points are now sold by dealers in hardware, and every smith knows how they are shaped. There is, however, no need of buying these; every smith has old plowshares from which points can be cut, provided you don't use an old share too much worn. The points sold are cut with the intention that most of the point is to be placed on top of the plow point. This is all right in some instances, while it is wrong in others. When you cut a piece for a point make it the same shape at both ends. Now, when a plow needs the most of the point on top bend the end to be on top longer than the end to go underneath, and vice versa, when the point wants to be heaviest on the bottom side. I hold that in ordinary cases the most of the point should be on the bottom side. If it is it will wear better and keep in the ground longer, for as soon as the point is worn off underneath it comes out of the ground.

Don't monkey with old mower sections or anything like them for points, for, although the material is good, it is not the quality alone but also the quantity that

JAPANESE PLOW

goes to make up a good point. It takes only a few hours' plowing to wear off a section from the extreme point of the share, and then there is only the iron of the plow point left to wear against, and your time spent for such a point is lost. Another thing, it takes just as much time to put on such a point as it does to put on a good one for which you charge the regular price.

In putting on a point of thin material you must go unusually slow, or you will burn the steel before the plow point is hot.

Smiths, as a rule, draw out a round back point. They seem to be afraid of coming down on the point with the hammer for fear it will spring the point towards the land. This can be remedied by using a wooden block for anvil. Then you can set the point back without battering the edge of the share. The

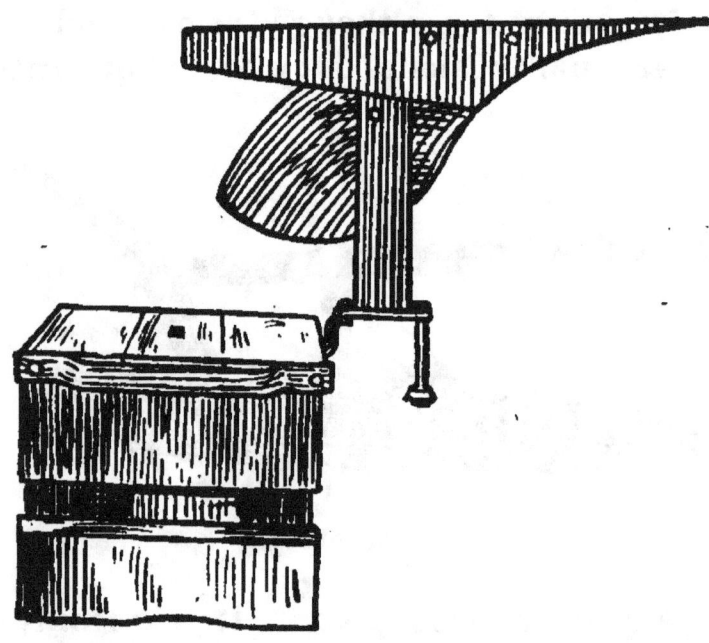

BENCH FOR HOLDING PLOWS WITHOUT BEAM

suck of a point should be one-eighth of an inch. Don't split the steel of the point of a share open and wedge a point in. Make one long enough to reach around the point, say from 8 to 10 inches long, and you will have a good substantial job. There is too much experimenting in putting on points yet, but the method just described is the only good one.

HOW TO SHARPEN A PLOWSHARE

If the share to be sharpened is a hardened share, and it is the first time it is sharpened, then be careful not to heat it too far towards the joint, so as to leave the temper as much the same as possible. For my part, I never follow this rule. I heat it as much as is needed to draw it out good, and then harden it over again. But beginners can sharpen a new share once without hardening it over, if the temper is not entirely out of the share. To sharpen a share without springing it some is an impossibility. No device will prevent this, and the only way to set it right is to heat it all over. In sharpening a share it is drawn out on one side, and it is natural that that side is made longer, and as a result the share must warp. In a circular saw it takes only a couple of blows on one side to get it out of shape; then what else can we expect in a plowshare, when all the hammering is done on one side?

Some smiths turn the bottom side of the share up and hammer on that side, but this is wrong; first, because in so doing you unshape the share; second, the scales on the anvil will mark the face of the share just as bad as the hammer, so nothing is gained by this. Place the share on the anvil, face up, and use a hammer with a big round face, and when you get used to this, the best result is obtained. D n't draw the edge out too thin. There is no need of a thin edge on a plow that has to cut gravel and snags, but for sod breaking a thin edge is wanted, and the smith has to use his best judgment even in such a case.

HOW TO PUT ON A HEEL

Cut a piece of steel about eight inches long, **three** inches wide on one end, and pointed down to a sharp point on the other. Draw out one side thin to nothing. Next, draw out the heel of the share. Now place the heel piece on the bottom side of the share, and hold it in place with a pair of tongs and tong rings. Take the first heat at the pointed end of the piece, next heat at the heel, share down, then turn the share over, heel down; go slow, use borax freely, and place a little steel borings between the heel piece and the share. After a little practice almost any smith ought to be able to put on a heel, while now it is only a few smiths that can do it. I never put on a heel yet but the owner of the plow would tell me that other smiths tell him it cannot be done. When welded good be sure to get the right shape in the share. Grind and polish carefully, as the dirt is inclined to stick to the share in this place more easily than in any other.

HOW TO REPAIR A FLOPPING PLOW

When a plow is flopping or going everywhere so that the owner don't know what is the matter the fault should be looked for first in the beam. If the beam is loose the plow will not run steady, but the reason for this trouble, in most cases, is in the share. If the point has too little "suction," and the edge of the share is too much rolling the plow generally acts this way.

To remedy this, sharpen the share, set the point down, and the edge of the lay from the point all the way back to the heel, and the plow will work right.

HOW TO SET A PLOW RIGHT THAT TIPS ON ONE SIDE

If a plow is inclined to fall over on the right handle, the fault is in the share. The share in such a case has too much suction along the edge. Heat the whole share and roll the edge of it up and the plow will work all right.

If a plow tips over on the left side handle, the share in such a case is too much rolled up. Heat it all over and set the edge down to give it more suction.

WHEN A PLOW RUNS TOO DEEP

There are two reasons for a plow running too deep: 1, If the beam is more than fourteen inches high from the floor up to the lower side of it, then the beam should be heated over a place as far back as possible, and the same set down to its proper place. 2, If the point of the share has too much suction the plow will also run too deep. The right suction to give a plowshare is from $\frac{1}{8}$ to $\frac{3}{16}$ of an inch. If a plow don't run deep enough with this much as a draw, there must be something else out of shape; or, if it goes too deep, the fault must be looked for in the beam or in the tugs with small-sized horses.   The point of a share should never be bent upwards in order to prevent the plow from going too deep.   Set the share right, and if the plow then goes out of its proper way the fault must be found somewhere else.

WHEN A PLOW TAKES TOO MUCH LAND

If a 14-inch plow takes too much land the fault is either in the point of the share or in the beam.   The point of a share should stand one-eighth of an inch to land, and the beam should stand about three inches to the right.   This will be right for a 14-inch plow and two horses.   If for a 16-inch plow and three horses, the beam should be in line with the landside.

HOW TO FIX A GANG PLOW THAT RUNS ON ITS NOSE

When a gang or sulky plow runs on its nose and shoves itself through the dirt, the fault is with the share or in the beam.   In most cases this fault is a set back beam, but it might also be the result of a badly-bent-down and out-of-shape landside point.   If it is in the beam, take it out and heat it in the arch, then bend it forward until the plow has the right shape, and it will run right.

HOW TO HARDEN A MOULDBOARD

To harden a mouldboard is no easy job in a blacksmith's forge, and it is no use trying this in a portable forge, because there is not room enough for the fire required for this purpose. First, dig the firepot out clean, then make a charcoal fire of two bushels of this coal, have some dry basswood or wood like it, and when the charcoal begins to get red all over then pile the wood on the outside corners of the fire. Heat the point of the mouldboard first, because this being shinned, it is thicker and must be heated first or it will not be hot enough; then hold the mouldboard on the fire and pile the wood and hot coal on top of it. Keep it only until red hot in the same place, then move it around, especially so that the edges get the force of the fire, or they will be yet cold while the center might be too hot.

HOW TO PATCH A MOULDBOARD

When the mouldboard is red hot all over sprinkle with prussiate of potash, and plunge into a barrel of ice or salt water. A mouldboard will stand a good heat if the heat is even; otherwise it will warp or crack. Another way to heat a mouldboard: if you have a boiler, then fill the fire place with wood and heat your mouldboard there. This will give you a very good heat. If it is a shinned mouldboard the point must be heated first in the forge, then place it under the boiler for heating. This must be done to

insure a good heat on the point, which is thicker than
the mouldboard and therefore would not be hot enough
in the time the other parts get hot.

When a mouldboard is worn out on the point a patch
can be put on, if the mouldboard is not too much worn
otherwise. Cut a piece of soft center steel to fit over
the part to be repaired. Draw this piece out thin
where it is to be welded to face of mouldboard. Hold

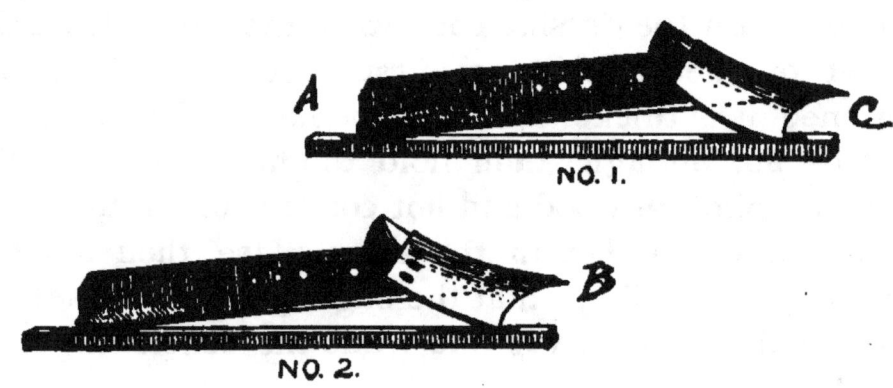

FIG. 11

this piece in position while taking the first weld, with
a pair of tongs. Weld the point first, then the edges,
last the center. The patch should be welded to face
of mouldboard. When the last weld is taken place the
mouldboard face up, with some live coal over it, in the
fire; use borax freely, and, when ready to weld, weld
the patch while the mouldboard is in the fire, using a
⅜ rod of round iron as a hammer with one end of it
bent for this purpose. When the patch is thus welded
in its thinnest place then take it out and weld on the
anvil. In heating for the weld never place the patch
down towards the tuyer, for there the blast will make

it scale, and it will never weld this way. Remember this in all kinds of welding.

Figure 11A represents two shares. No. 1 represents a share set for spring plowing, when the ground is soft. Notice the heel of the share following the square for about one inch at c, while the heel in No. 2 rests with the extreme edge on the square, and is set for fall plowing, when the ground is hard. The line between a and b shows the suction at d, which is not more than an eighth of an inch. Breaking plows and large plows which are run shallow should have a wide bearing at c. In breaking plows the heel will sometimes have to be rolled up a little at this place.

"The reason most men do not accomplish more is because they do not attempt more."

CHAPTER VII

MOWER SECTIONS

N filling a sickle bar there are two ways to remove the old sections. One way is to punch the rivets out, but in every case where the back of the section sticks out over the sickle bar they can be removed easier in this way: Just open the vise enough to receive the section, then strike with the hammer on the back of the section, and this blow will cut the rivets off. You can cut out ten to one by this method to any other.

Sometimes the sickle bar is bent out of shape in the fitting. To straighten it place the sickle on the anvil, sections down; now strike with the hammer so that it will touch the bar only on one half of its face, the blow to be on the inner side of the curve.

BABBITING

When a box is to be babbited the first thing to do is to clean the box If it can be placed over the fire the old babbit with melt out easily. If the box cannot be held over the fire, then chisel the old babbit out. At each end of the box there is a ridge to hold the babbit in the box; that is, in cast iron boxes. On top of this ridge place a strip of leather as thick as you want the babbit to be. This done, place the shaft in the box. Pour the babbit in level with the box. Be careful about having the box dry; if any dampness is in the box the babbit will explode. Now place a thick paper on each side of the box and put on the top box, with the bolts in to hold it in place tight, then close up at the ends with putty. In some cases it is best to heat the box a little, for if the box is cold and there is little room for the babbit it will cool off before it can float around. In such a case the boxes should be warm and the babbit heated to a red heat. Now pour the babbit in through the oil hole.

In cases where there are wooden boxes, and the babbit is to reach out against the collars, the shaft must be elevated or hung on pieces of boards on each side with notches in for the shaft to rest in. Use putty to fill up and make tight, so that the babbit must stay where wanted. For slow motion babbit with a less-cooling percentage (tin); for high speed, more-cooling (tin). Grooves may be cut in the bottom box for oil. When a shaft is to be babbited all around in a solid box the shaft is inclined to stick in the babbit. To prevent this smoke the shaft a little and have it warm. When cool it will come out all right. Or wind thin paper around the shaft, the paper to be tied with strings to the shaft.

ANNEALING

By gradually heating and cooling steel will be softened, brittleness reduced, and flexibility increased. In this state steel is tough and easiest drilled or filed. Tool steel is sometimes too hard to drill or file without first annealing it; and the best way to do this is to slowly heat to a red heat, then bury the steel in the cinders and let it cool slowly. To heat and let the steel cool exposed to the air will do no good, as it cools off too quick, and when cool the steel is as hard as ever. This is air temper.

HOW TO REPAIR BROKEN COGS

Cogs can be inserted in a cogwheel in different ways. If the rim of the wheel is thick enough a cog can be dovetailed in. That is, cut a slot in the rim from the root of the cog down, this slot to be wider at the bottom. Prepare a cog the exact size of the cogs, but just as much deeper as the slot. Before you drive this cog in, cut out a chip on each end of the slot, and when the cog is driven in you can clinch the ends where you cut out. This will make a strong cog, and if properly made will never get loose.

Another way: If the rim is thin, then make a cog with a shank on, or a bolt cog. If the rim is wide make two bolts. The cog can be either riveted or fastened with nuts. If only one shank is made, the same must be square up at the cog, or the cog will turn and cause a breakdown. But a shallow slot can be cut in the rim to receive and hold the cog, and then a bolt shank will hold it in place, whether the shank is round or square.

HOW TO RESTORE OVERHEATED STEEL

If steel has been burnt the best thing to do is to throw it in the scraps; but if overheated it can be improved. Heat to a low red heat, and hammer lightly and cool off in salt water, while yet hot enough to be of a brown color. Repeat this a half a dozen times, and the steel will be greatly bettered. Of course, this is only in cases when a tool or something like it has been overheated which cannot be thrown away without loss. By this simple method I have restored tools overheated by ignorant smiths, and in some cases the owner would declare that it was "better than ever."

HOW TO DRESS AND HARDEN STONE HAMMERS

Care must be taken in heating stone hammers not to overheat them. Dress the hammer so that the edges are a little higher than the center, thus making a slight curve. A hammer dressed this way will cut better and stay sharp longer than if the face is level. Dress both ends before hardening, then harden face end first. Heat to a red heat, and cool off in cold water about one inch up, let the temper return to half an inch from the face, that is, draw the temper as much as you can without changing the temper at the face. There it should be as hard as you can make it. When heating the peen end keep a wet rag over the face to prevent it from becoming hot.. This end should not be tempered quite as hard as the face.

HOW TO DRILL CHILLED CAST IRON

Chilled cast iron can be easily drilled if properly annealed, but it cannot be annealed simply by heating and slowly cooling. Heat the iron to a red heat and place it over the anvil in a level position; place a piece of brimstone just where the hole is to be drilled, and let it soak in. If it is a thick article place a piece on each side over the hole, as it will better penetrate and soften the iron. Next, heat it again until red, then bury it in the cinders, and let it cool slowly. To heat and anneal chilled iron is of no avail unless it is allowed to remain hot for hours. Chilled iron will, if heated and allowed to cool quick, retain its hardness. The only way to anneal is to let it remain in the fire for hours. Brimstone will help considerably, but even with that it is best to let cool as slowly as your time will admit.

HOW TO DRILL HARD STEEL

First, make your drill of good steel, oval in form, and a little heavier than usual on point, and temper as hard as it will without drawing the temper, the heat to be a low red cherry. Diluted muriatic acid is a good thing to roughen the surface with where you want the hole. Use kerosene instead of oil, or turpentine. The pressure on the drill should be steady so that it will cut right along as it is hard to start again if it stops cutting, but if it does, again use diluted muriatic acid. The hole should be cleaned after the use of the acid.

FACTS ABOUT STEEL

I have repeatedly warned against overheating steel. Don't heat too fast, for if it is a piece of a large dimension the outside corners will be burnt, while the bar is yet too cool inside to be worked. Don't let steel remain for any length of time in the fire at a high heat, for both steel and iron will then become brittle. This is supposed by some to be due to the formation of oxide disseminated through the mass of the metal, but many others believe that a more or less crystalline structure is set up under the influence of a softening heat, and is the sole cause of the diminution in strength and tenacity. The fiber of the steel is spoiled through overheating; this can, to some extent, be remedied by heavy forging if it is a heavy bar.

Steel is harder to weld than iron, because it contains less cinders and slag, which will produce a fusible fluid in iron that will make it weld without trouble. Steel contains from 2 to 25 per cent carbon, and varies in quality according to the per cent of carbon, and it is claimed that there are twenty different kinds of steel. To blacksmiths only a few kinds are known, and the sturdy smith discards both "physical tests and chemical analysis," and he thinks he knows just as much as do those who write volumes about these tests.

To weld tool steel, or steel of a high per cent of carbon, borax must be used freely to prevent burning and promote fusing. Steel with less carbon, or what smiths call "soft steel," "sleigh steel," should be welded with sand only. This soft steel stands a higher heat than the harder kinds.

Good tool steel will break easy when cold if it is cut into a little with a cold chisel all around, and the bar then placed with the cut over the hole in the anvil, the helper striking directly over the hole. If it is good steel it will break easy, and the broken ends are fine grain, of a light color. If it shows glistening or glittering qualities it is a bad sign.

Good steel will crumble under the hammer when white hot.

To test steel draw out to a sharp point, heat to a red heat, cool in salt water; if it cuts glass it is a steel of high hardening quality.

For armor piercing, frogs, tiles, safes, and crushing machinery, alloy steel is used. This steel contains chromium, manganese or nickel, which renders it intensely hard. Tungsten is another alloy that is used in iron-cutting tools, because it does not lose its hardness by friction. Smiths should know more about steel than they do, and we would have steel to suit every need. As it is now, any poor stuff is sent to the smith. The same can be said of iron. The American wrought iron is the poorest iron that ever got the name of iron, but there are thousands of smiths using this stuff with great difficulty without ever a word said as a protest against the manufacture of the rotten material.

We often get iron that is too poor to bend hot without breaking. Let us register a kick, and if that has no effect let us try to abolish the tariff, and there will be good iron manufactured in this country, or the Swedish and Norwegian iron will be used. But the result will be the same with iron as with the matches: the American manufactories will make good iron when they have to. We get iron and steel that is both "cold-shot and hot-shot." The former breaks easy when cold, the latter when hot. We have meat and wheat inspectors; where is the iron inspector? Farmers know enough to ask for protection, but blacksmiths will never say a word. They use the cold-shot or hot-shot iron, and when they have spent half a day in completing a little intricate work it breaks in their hands because of iron that is either cold or hot shot. 'nsist' on good iron, and the steel will also be good. Deduct a little every year from the amount due your jobber for poor iron, and you can be sure if this is done by a few thousand smiths it will have effect.

HOW TO WELD CAST IRON

Strictly speaking, there is no such thing as welding cast iron. The best that can be done is to melt it together; but this is simply accidental work, and when done don't amount to anything. Still, I have never met a blacksmith yet who could not weld cast iron, but, at the same time, I have yet to meet the man that can do it; and I will give twenty-five dollars to the smith that will give me a receipt for welding cast-iron shoes that will be useful when welded. All receipts I have seen for this purpose are simply bosh.

Malleable iron is a different thing. Many smiths weld malleable iron and think it is cast iron. "The wish," in such a case, "is the father of the thought," but to weld malleable iron is not more difficult than to weld soft steel. Malleable iron when good, and steel when soft, are about the same thing. I would therefore advise smiths to spend no time in welding cast iron. Nothing will be gained even if you should succeed in sticking it enough to hang together. It will in most cases be useless, because it will not be of the same shape as before.

F a pump handle is broken use rock salt and powdered glass as a welding compound. Stick the ends together in the fire. When they are about ready to melt tap lightly on one end while your helper holds the other end steady. In one case out of a hundred it will stick enough to hang together. If you have nothing else to do this will be a nice thing to kill time with

CASE HARDENING

Iron and steel may be case hardened with either of the following compounds: Prussiate of potash, sal-ammoniac of equal parts. Heat the iron red hot and sprinkle it with this compound, then heat again and sprinkle, and plunge it while yet hot in a bath of salt water.

Another: Cyanide of potassium; grind it into a fine powder and sprinkle over the iron while red hot, and plunge into a bath of salt water. This powder will coagulate if it is held against the fire so it gets warm. Be careful with this powder, as it is a strong poison. It is the best thing that I have ever tried for case hardening iron. It will case harden the softest iron so that it cannot be touched with any tool. It is also good for plows, especially where it is hard to make a plow scour. The only objection is the price, as it costs more than prussiate of potash or other hardening compounds.

HOW TO HARDEN SPRINGS

Heat to a heat that will be discerned in the dark as a low red heat. Plunge into a bath of lukewarm water. Such a heat cannot be noticed in a light sunny day, but it is just the heat required. Of course, it is the smith with practice who succeeds, as with everything else.

Another way: Heat to a low red heat and bury the spring in cold sand. Another: Heat to a low red heat in the dark, and cool in oil.

TO MAKE STEEL AND IRON AS WHITE AS SILVER

Take 1 pound of ashes from white ash bark, dissolve in soft water. Heat your iron red, and cool in this solution, and the iron will turn white as silver.

TO MEND BROKEN SAWS

Silver, 15 parts; copper, 2 parts. These should be filed into powder and mixed. Now place your saw level with the broken ends tight up against each other; put a little of the mixture along the seam, and cover with powdered charcoal; with a spirit lamp and a blowpipe melt the mixture, then with the hammer set the joint smooth.

TO MEND A BAND SAW

If a band saw is broken file the ends bevel, and lap one end over the other far enough to take up one tooth; place the saw in such a position that the saw will be straight when mended; use silver, copper and brass; file into a fine powder; place this over the joint and cover with borax. Now heat two irons one inch square, or a pair of heavy tongs, and place one on each side of the joint, and when the powdered metal is melted have a pair of tongs ready to take hold over the joint with while it cools. File off and smooth the sides, not leaving the blade any thicker than in other places.

TO WRITE YOUR NAME ON STEEL

TAKE of nitric acid 4 ozs.; muriatic acid, ½ oz. Mix together. Now cover the place you wish to write on with beeswax, the beeswax to be warm when applied. When it is cold, write your name with a sharp instrument. Be sure to write so that the steel is discernible in the name. Now apply the mixture with a feather, well filling each letter. Let the mixture remain about five minutes or more, according to the depth desired; then wash off the acid; water will stop the process of the same. When the wax is removed, the inscription is plain.

"The man who confesses his ignorance is on the road to wisdom."

CHAPTER VIII

HOW TO PATCH A BOILER

By H. MOEN, MACHINIST, CRESCO, IOWA.

WHEN the leak or weak place in the boiler is found, take a ripping chisel and cut out all of the weak, thin and cracked parts. This done, make the patch. The patch must be large, not less than an inch lap on all sides, but if double rows of rivets are wanted the lap should be two inches on all sides. Bevel or scrape the patch on all edges to allow calking. The bolt holes should be about two inches apart and countersunk for patch bolts, if patch bolts are used. Next, drill two holes in the boiler shell, one on each side of the patch, and put in the bolts. These bolts should be put in to stay and hold the patch in position while the rest of the holes are drilled and bolted. When the bolts are all in, take your wrench and tighten the bolts one after the other, harder and harder, striking at the same time on the patch around its edges. At last strike light on the bolt heads when you tighten and draw the bolt until its head breaks off. These bolts are made for this purpose and in such a shape that the head will break at a high strain. This done use the calking iron all around the patch.

The patch should be put on the inside of the boiler, especially if on the bottom of a horizontal boiler. If the patch is put on the outside in this place the sediment or solid matter which the water contains will quickly fill up over the patch and there is danger of overheating the boiler and an explosion may follow.

HOW TO PUT IN FLUES

The tools necessary to retube an old boiler are, first, a good expander of the proper size; a roller expander preferred; a crow foot or calking iron, made from good tool steel. A cutting-off tool can be made to do very

TUBE TO BE WELDED

good service, in the following manner: Take a piece of steel, say ½ x 1¼, about ten inches long. Draw one end out to a sharp point and bend to a right angle of a length just enough to let it pass inside of the flue to be cut. A gas pipe can be used for a handle. In cutting the flues set this tool just inside the flue sheet and press down on the handle. If this tool is properly made it cuts the old flues out with ease. After both ends have been cut the flues will come out.

Next, cut the tubes about ⅜ of an inch longer than the flue sheet. After the tubes are cut the proper length, and placed in the boiler, expand the same in both ends with a flue expander. After the flues are expanded until they fit the holes solid, turn them over with the peen of a hammer to make them bell shaped. Now take a crows-foot, or calking tool, and turn the

TUBE EXPANDER

ends in a uniform head and tight all around. If the flues should leak, and there is water on the boiler take a boiler expander and tighten them up. But never attempt to tighten a flue with the hammer if there is water on the boiler.

HOW TO WELD FLUES

In welding flues or putting new tips on old flues, you must find out how far the old tubes are damaged, and cut that part off. Next clean the scales off in a tumbling box; if you have none, with an old rasp.

Now take a piece of tubing the size of the old, and scarf the ends down thin, the new tube to go over the old and drive them together. In welding a rest can be made in the forge to push the tube against while welding, to prevent the pieces from pulling apart. A three-eighths rod, with thread on one end and a head

on the other, run through the flue will be found handy
for holding the pipes or flues together. In welding
these together don't take them out of the fire and
strike with a hammer, but take a rod ⅜-inch round,
and bend one end to serve as a hammer. Strike with
this hammer lightly over the lap, at the same time
turning the flue around in the fire. Use borax to prevent the flue from scaling and burning.

FOAMING IN BOILERS

There are many reasons for foaming in boilers, but
the chief reason is dirty water. In some cases it is
imperfect construction of boiler, such as insufficient
room for the steam and a too small steam pipe or dome.
When a boiler is large enough for the steam and clean
water is used there is no danger of foaming. When
more water is evaporated than there is steam room or
heating surface for, then the boiler will foam. When
a boiler is overworked more steam than its capacity
will admit is required, and the engine is run at a high
speed, the steam will carry with it more water than
usual.

When a boiler foams shut the throttle partly to
check the outflow of steam and lessen the suction of
water, because the water is sucked up and follows the
sides of the dome up.

If the steam pipe in the dome sticks through the
flange a few inches the water will not escape so easy.
A boiler that is inclined to foam should not be filled

too full with dirty water; if it is it is best to blow off a little. Foul water can be cleaned by different methods before it enters the boiler, so as to prevent foaming and scaling.

BLOWING OUT THE BOILER

A boiler should not be blowed out under a high steam pressure, because the change is so sudden that it has a tendency to contract the iron, and if repeated often the boiler will leak. If it is done when there is brickwork around the boiler and the same is hot it will in a short time ruin the boiler. In such a case the boiler should not be blowed out for hours after you have ceased firing.

'A trained man will make his life tell: without training we are left on a sea of luck, where thousands go down while one meets with success."—Garfield.

CHAPTER IX

THE HORSE

HE horse must have been one of the first animals subjected to the use of man, but there is no record made of it before the time of Joseph, during the great famine in Egypt, when Joseph exchanged bread for horses. During the exodus horses were used more extensively, and in consequent wars we find the horse used especially by great men and heroes. This noble animal has always been held in high esteem by civilized people. In wars and journeys and exploits, as well as for transports, the horse is of immeasurable value. No people cared for and loved this animal as did the Arabs. The care and breeding of horses was their main occupation, therefore their horses were noted for intelligence, high speed and endurance. The English and American thoroughbred has an infusion of blood of the Arab horse, which has set the price on these animals. Pedigrees were first established by the Arabs.

Each country has its own breed of horses. Horses

of a cold climate are smaller in size, as also are the horses of the tropics. The best horses are found in the temperate zone. In Germany the horses are large, well formed and strong. Norway and Sweden have a

FIG. 3.

race of little horses, and not until a few years ago did the people of these countries know anything about pedigrees; their horses are spirited and stronger in porportion to the size than any other race of horses. In Sweden and Norway the farmer, with wife and children, will walk many miles Sunday to church, while the horses roam in the pasture or stand in the stable.   Some farmers will not hire out their team for money.   The horses of these countries are better taken

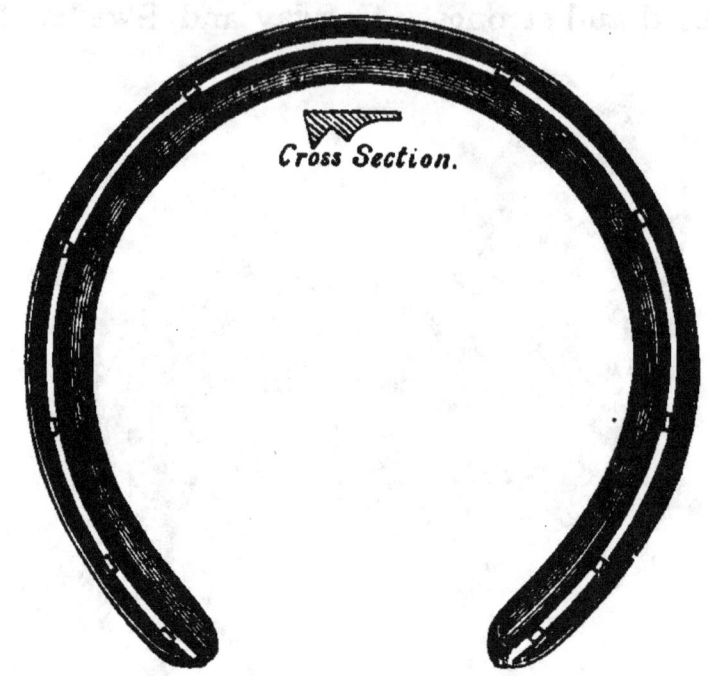

FIG. 51.—TOE AND SIDE WEIGHT AND PLAIN RACING PLATES, AS
MANUFACTURED BY BRYDEN HORSE SHOE CO.

care of than anywhere else, of course with the exception of American race horses.

HORSE-SHOEING

The horse in a wild state needs no shoes, the wear and tear that the feet are subjected to while the horse is hunting for his food in a wild country on soft meadows, is just right to keep the hoofs down in a normal condition.   But when the horse is in bondage and must serve as a burden-carrying animal, traveling on hard

roads or paved streets, the horse must be shod to prevent a foot wear which nature cannot recuperate.
Horseshoes were first made of iron in 480 A. D.
Before that time, and even after, horseshoes have
been made of leather and other materials.

**FIG. 52.—TOE AND SIDE WEIGHT AND PLAIN RACING PLATES, AS
MANUFACTURED BY BRYDEN HORSE SHOE CO.**

ANATOMY

It is necessary in order to be a successful horse-shoer
to know something about the anatomical construction
of the feet and legs of the horse. Of course, any little
boy can learn the names of the bones and tendons in
a horse's foot in an hour, but this does not make a
horse-shoer out of him. No board of examiners should

allow any horse-shoer to pass an examination merely because he can answer the questions put to him in regard to the anatomy of the horse, for as I have said before, these names are easily learned, but practical

FIG. 53.—TOE AND SIDE WEIGHT AND PLAIN RACING PLATES, AS MANUFACTURED BY BRYDEN HORSE SHOE CO.

horse-shoeing is not learned in hours; it takes years of study and practice.

It is not my intention to treat on this subject. I could not; first, because there is not room for such a discourse, second, there are numerous books on the subject better than I could write, available to every horse-shoer. I shall only give a few names of such parts of the anatomy as is essential to know. What the horse-shoer wants to know is the parts of the foot connected with the hoof, as his work is confined solely to the foot.

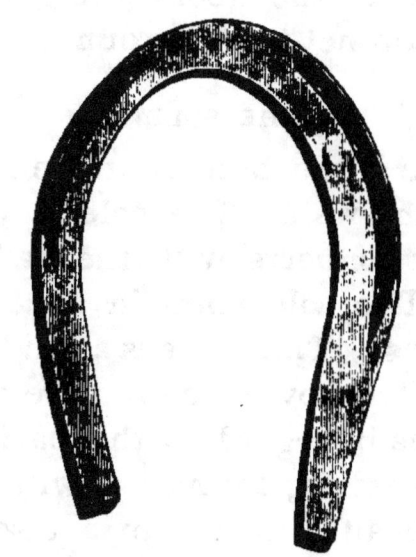

FIG. 54. —TOE AND SIDE WEIGHT AND PLAIN RACING PLATES. AS MANUFACTURED BY BRYDEN HORSE SHOE CO.

THE WALL

The wall or crust is the horny sheath incasing the end of the foot, in the front and on the sides from the coronet to the ground. It is through this crust the nail is driven, and it is upon this crust the shoe rests. In front it is deepest, towards the quarter and heel it becomes thinner. It is of equal thickness from the upper end to the ground (from top to bottom). The white corored wall is the poorest, while the iron colored wall is the toughest. The growth of the wall is different at different ages. It grows more in a young horse and colt than in an old horse; in a healthy foot and soft, than in a diseased foot and hard. In a young horse the hoof will grow about three inches in a year and even more, while it grows less in an old horse. The wall is fibrous, the fibers going parallel to each other from the coronet to the ground.

THE SOLE

The horny sole is the bottom of the foot. This sole is fibrous like the wall. The sole is thickest at the border, where it connects with the wall, and thinnest at the center. The sole when in a healthy condition scales off in flakes. This scale is a guide to the farrier whereby he can tell how much to pare off. There are different opinions in regard to the paring of the sole, but that is unnecessary, for nature will tell how much to cut off in a healthy foot. In a diseased foot it is different; then the horse-shoer must use his own good judgment. It is, however, in very few cases that the shoer needs to do more than just clean the sole. Nature does the scaling off, or paring business, better than any farrier.

THE FROG

The frog is situated at the heel and back part of the hoof, within the bars; the point extending towards the center of the sole, its base filling up the space left between the inflection of the wall. This body is also fibrous. The frog is very elastic and is evidently designed for contact with the ground, and for the prevention of jars injurious to the limbs.

CORONET

Coronet is the name of the upper margin of the foot, the place where the hair ceases and the horny hoof begins.

THE QUARTER

The quarter means a place at the bottom of the wall, say, about one-third the length from the heel towards the toe.

THE BARS

By the bars we mean the horny walls on each side of the frog, commencing at the heel of the wall and extending towards the point of the frog.

Any blacksmith or horse-shoer desiring to study more thoroughly the anatomy of the horse should procure a book treating on this subject.

HOW TO MAKE THE SHOE

It is only in exceptional cases that the shoer turns or makes a shoe. The shoes are now already shaped, creased and partly punched, so all that is needed is to weld on the toe calk and shape the heel calks.

Heat the shoe at the toe first, and when hot bend the heels together a little. This is done because the shoes will spread when the toe calk is welded on, and the shoe should not be too wide on the toe, as is mostly the case. If the shoe is narrow at the toe it is easier to fit the same to the foot and get the shoe to fill out on the toe. Many smiths cut too much off from the toe. Before the toe calk is driven onto the shoe bend it a little so as to give it the same curve the shoe has, and the corners of the calk will not stick out over the edge of the shoe. Now place the shoe in the fire, calk up. Heat to a good low welding heat, and use sand for welding compound. Don't take the shoe out of the fire to dip it in the sand, as most shoers do, for you will then cool it off by digging in the cold sand, of which you will get too much on the inner side of the calk. The same will, if allowed to stay, make the calk look rough. You will also have to make a new place for the shoe in the fire, which will take up a good deal of time, as the new place is not at once so hot as the place from which the shoe was taken; besides this, you might tear the calk off and lose it. When hot give a couple of good blows on the calk and then draw it out. Don't hold the heels of the shoe too close to the anvil when you draw out the calk, for if you do the calk will stand under, and it should be at a right angle with the shoe. Do not draw it out too long, as is mostly done. Punch the hole from the upper side first. Many first-class horse-shoers punch only from that side, while most shoers punch from both sides.

There is no need of heating the shoe for punching the holes. Punch the holes next to the heel first, for if you punch the holes next to the toe when the shoe is hot, the punch will be hot, upset and bent. If it is a large shoe, punch only two holes on each side for the toe calk heat. These holes to be the holes next to the toe when the shoe is hot, and then punch the other two when you draw out the heel calks, and the shoe is hot at the heel. The heel calks should be as short as you can make them; and so should the toe calks. I know but a few horse-shoers that are able to weld on a toe calk good. The reason for their inability is lack of experience in general blacksmithing. Most shoers know not how to make a fire to weld in. They are too stingy about the coal; try to weld in dirt and cinders, with a low fire, the shoe almost touching the tuyer iron. I advise all horse-shoers to read my article about the fire.

I have made a hammer specially for horse-shoeing with a peen different from other hammers. With this hammer the beginner will have no trouble in drawing out the calks. See Figure 8, No. 8. The hammers as now used by most smiths are short and clumsy; they interfere too much with the air, and give a bump instead of a sharp cutting blow that will stick to the calk.

The shoe should be so shaped at the heel as to give plenty of room for the frog; the heels to be spread out as wide as possible. This is important, for if the shoe is wide between the heels the horse will stand more firm, and it will be to him a comfortable shoe. The shoe should not be wider between the calks at the expense of same, as is done by some shoers, for this is only a half calk, and the heel is no wider. The shoe should not be fitted to the foot when hot, as it will injure the hoof if it is burned to the foot.

HOW TO PREPARE THE FOOT FOR THE SHOE

The foot should be level, no matter what the fault is with the horse. The hoof should not be cut down more than the loose scales will allow. In a healthy condition this scale is a guide. When the foot is diseased it is different, and the shoer must use his own judgment.

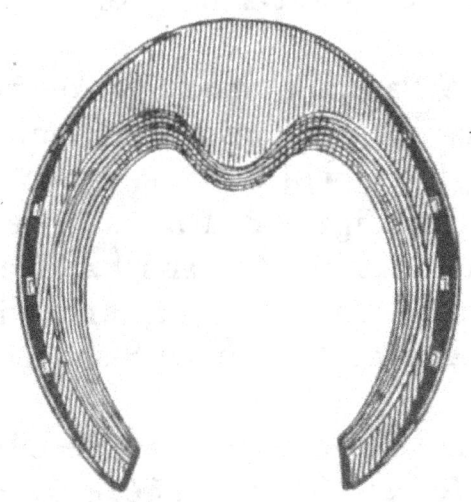

The frog never grows too large. It should never be trimmed more than just to remove any loose scales.

The frog in its functions is very important to the well-being of the foot. In the unshod foot it projects beyond the level of the sole, always in contact with the ground; it obviates concussion; supports the tendons; prevents falls and contraction. The bars are also of importance, bracing the hoof, and should never be cut down as has been the practice for centuries by ignorant horse-shoers.

FORGING

Forging or overreaching is a bad habit, and a horse with this fault is now very valuable. This habit can be overcome by shoeing; but it will not be done by making the shoes short on the heel in front and short in the toe behind. Never try this foolish method.

To overcome forging the shoer should know what forging is. It is this: The horse breaks over with his

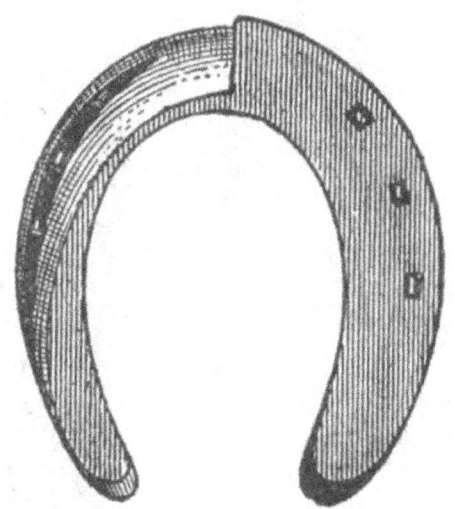

hind feet quicker than he breaks over with the front feet; in other words, he has more action behind than in front, and the result is that the hind feet strike the front feet before they can get out of the way, often cutting the quarters badly, giving rise to quarter cracks and horny patches over the heel.

Some writers make a difference between forging and overreaching, but the cause of the trouble is the same —too much action behind in proportion to the front; and the remedy is the same—retard the action behind,

increase it in front. There are different ideas about
the remedy for this fault.

One method is to shoe heavy forward and light behind, but this is in my judgment a poor idea, although
it might help in some cases. Another way is to shoe

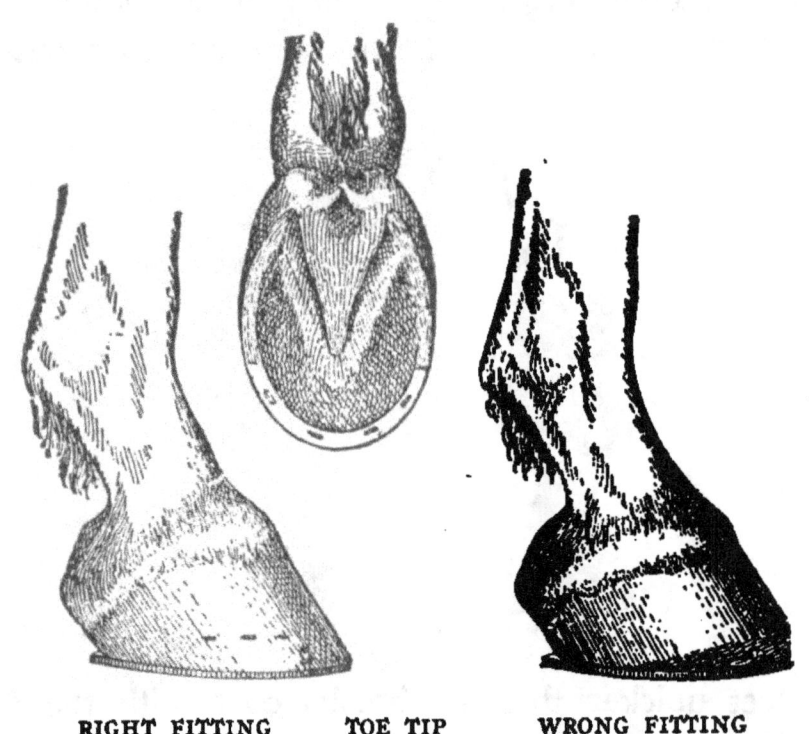

RIGHT FITTING   TOE TIP   WRONG FITTING

with side weight on the outer side behind, but it is not
safe, because it is difficult to get a horse to throw the
foot out to one side enough so as to pass by the front
foot except in a high trot.

The best way to shoe a forger or overreaching
horse is to make a shoe for front of medium heft, not
longer than just what is needed. The toe calk should

be at the inner web of the shoe, or no toe calk at all, or, toe weight, to make the horse reach farther.

It will sometimes be found that the hind foot is shorter than the front foot. To find this out, measure from the coronet to the end of the toe. The shorter the foot the quicker it breaks over. If it is found that the hind foot is shorter than the front foot, then the shoe should be made so that it will make up for this. Let the shoe stick out on the toe enough to make the foot of equal length with the front foot. It is well in any case of forging to make the hind shoe longer on the toe. If the hind shoe is back on the foot, as is often done, it will only make the horse forge all the more, for it will increase action behind, the horse breaks over quicker, and strikes the front foot before it is out of the way. Set the shoe forward as far as possible, and make long heels. The longer the shoe is behind the longer it takes to raise the foot and break over.

Clack forging is meant by the habit of clacking the hind and fore shoes together. This kind of forging is not serious or harmful; it will only tend to wear off the toe of the hind foot and annoy the driver, possibly a little fatiguing to the horse.

The position of the feet at the time of the clack is different from that it is supposed to be. The toe of the hind feet is generally worn off, while no mark is made on the front feet. From this you will understand that the hind feet never touch the heel of the front feet, but the shoe. Just at the moment the fore foot is raised up enough on the heel to give room for the

hind foot to wedge in under it the hind foot comes
flying under the fore foot, and the toe of the hind foot

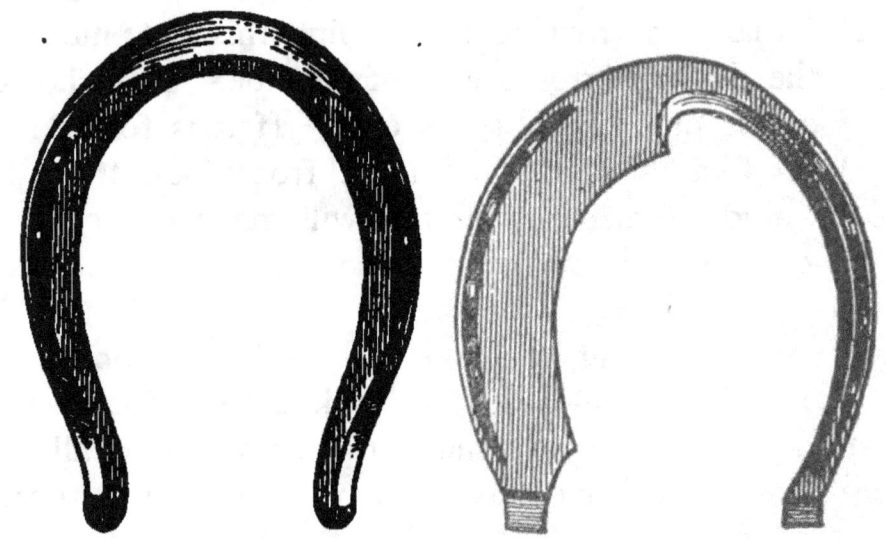

strikes the web of the toe on the front foot. This is
the reason no mark is seen on the front foot, while the
hind foot is badly worn off.

INTERFERING

Interfering is a bad fault in a horse. It is the effect
of a variety of causes. In interfering the horse brushes
the foot going forward against the other foot. Some
horses strike the knee, others above it, the shin or coronet, but in most cases the fetlock.

Colts seldom interfere before they are shod, but
then they sometimes interfere because the shoes are
too heavy. This trouble disappears as soon as the
colt is accustomed to carrying the shoes. Weakness is

the most common cause. Malformation of the fetlock is another cause. The turning in or out of the toes, giving a swinging motion to the feet, is also conducive to interfering.

The first thing to do is to apply a boot to the place that is brushed. Next, proceed to remove the cause by shoeing, or by feeding and rest in cases of weakness. Nothing is better than flesh to spread the legs

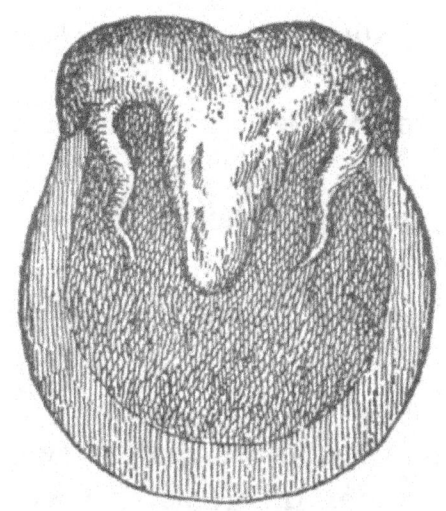

NATURAL FOOT

with. Some old horse-shoers in shoeing for interfering will turn the feet so as to turn the fetlock out. This is done by paring down the outside and leaving the inside strong. This is a bad way of shoeing for interfering, as it might ruin the horse. The foot should be leveled as level as it is possible. The inner side of the hoof should be scant; instead of being curved it should be almost straight, as the horse generally strikes with the side of the hoof or quarter. This is done to make a side-weight shoe, the side weight not to reach over the center of the shoe, but to be only on one side. Put the shoe on with the weight on the outer side. If the horse still interferes, give more side weight to the shoe, and make the heel on the outer side about one and one-quarter inch longer than the inside heel; give it an outward turn. This heel will prevent the horse from turning the heel in the way of the other foot when it goes by, so as not to strike the fetlock.

Properly made and applied, side weight will stop interfering almost every time. If the side weight is heavy enough it will throw the foot out, and the trouble is overcome.

There are only a few horse-shoers that have any practical experience in making side-weight shoes, which we understand from the articles in our trade journals.

Some horse-shoers in shoeing to stop interfering will make common shoes shorter than they ought to be and set them far in under the foot, so that the hoof on the inner side will stick out over the shoe a quarter of an inch. These they don't rasp off, and everybody knows that the hoof adheres to and rubs harder against the leg than the hard smooth shoe. But, foolish as it is, such shoers stick to their foolish ideas. I call all such fads faith cures.

The rule is to have the side weight on the outer side, while the exception is to have the side weight on the inner side of the foot. For old and poor horses ground feed and rest is better than any kind of shoes. It will give more strength and more flesh to spread the legs.

"Knowledge is of two kinds; we know a thing ourselves, or we know where we can find information upon it."—Dr. Samuel Johnson.

CHAPTER X

HOW TO SHOE A KNEESPRUNG OR KNUCKLER

NEESPRUNG is the result of disease that sometimes is brought about by bad shoeing. In a healthy leg the center of gravity is down through the center of the leg and out at the heels. This is changed in a case of kneesprung legs, giving the legs a bowed appearance. This trouble always comes on gradually; in some cases it will stop and never get worse, while in others it will keep on until it renders the horse useless. A horse with straight legs will sleep standing, but a knuckler cannot; he will fall as soon as he goes to sleep, on account of the center of gravity being thrown on a line forward of the suspensory ligaments. The cause of this trouble is sprain or injury to the back tendons of the legs; soreness of the feet, shins or joints. In old cases nothing can be done but just to relieve the strain a little by shoeing with a long shoe and high heel calks, with no toe calk. In cases not more than three months old clip the hair off the back tendons when there is any soreness, and shower them with cold water

149

several times a day for a week or two, and then turn the horse out for a long run in the pasture.

CONTRACTION

Contraction is in itself no original disease, except in a few cases. It is mostly the effect of some disease.

FOOT PREPARED FOR CHARTIER TIP　　FOOT SHOD WITH CHARTIER TIP

Contraction follows sprains of the tendons, corns, founder and navicular disease. When contraction is the result of a long-standing disease of the foot or leg it will be in only one of the feet, because the horse will rest the affected leg and stand most of the time on the healthy leg; thus the healthy foot receives more pressure than the diseased, and is spread out more; the foot becomes much uneven—they don't look like mates. This kind of contraction is generally the result of some chronic disease, but in most cases contraction is the result of shoeing and artificial living. Before the colt is shod his hoofs are large and open-heeled, the quarters are spread out wide, and the foot on the under side is shaped like a saucer. The reason of the colt's foot being so large is that he has been running on the green and moist turf, without shoes, and the feet have in walking in mud and dampness gathered so much moisture that they are growing and spreading at every

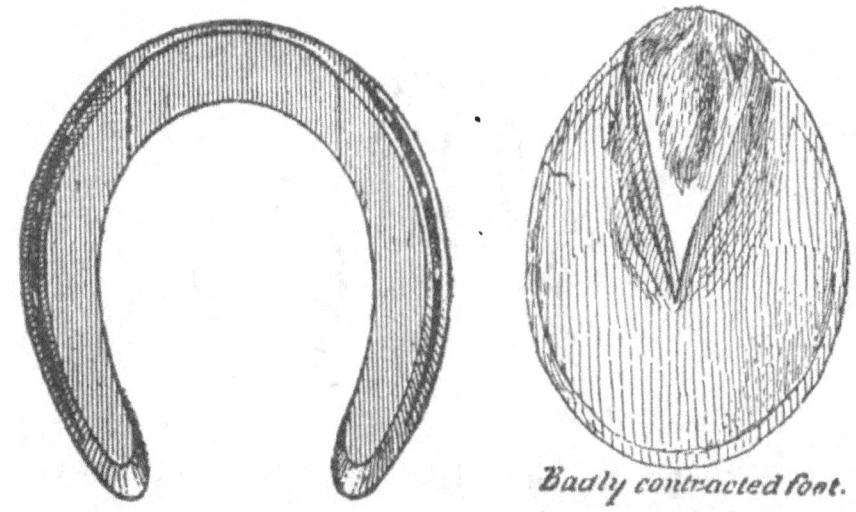

Badly contracted foot.

step. This is changed when the colt is shod and put on hard roads, or taken from the pasture and put on hard floors where the feet become hard and dried up. A strong high heeled foot is predisposed to contraction, while a low heeled flat foot is seldom afflicted with this trouble.

When contraction comes from bad shoeing or from standing on hard floors, pull the shoes off, pare down the foot as much as you can, leaving the frog as large as it is. Rub in some hoof ointment once a day at the

coronet and quarters, and turn the horse out in a wet
pasture.   But if the horse must be used on the road,
proceed to shoe as follows: First, ascertain if the frog
is hard or soft. If soft, put on a bar shoe with open
bar. I have invented a shoe for this purpose. See
Fig. 2, No. 1. The idea of shoeing with an endless
bar shoe is wrong. In most cases contraction is
brought on by letting the shoes stay on too long,
whereby the hoof has been compelled to grow down

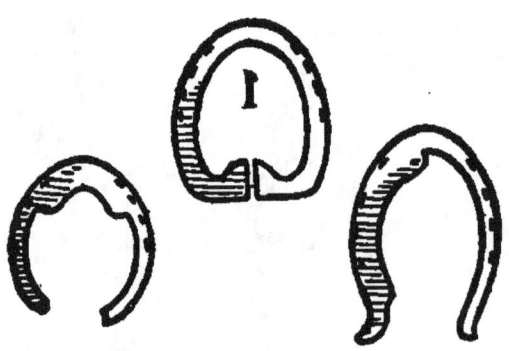

with the shape of the shoe. If an open shoe has
helped to bring on contraction, much more so will a
bar shoe, which will tie the hoof to the shoe with no
chance of spreading, no matter what frog pressure is
put on. Make the shoe as light as you can, with very
low or no calks; let the bar rest against the frog; keep
the hoofs moist with hoof ointment; use an open bar
shoe.

Make a low box and fill it with wet manure, mud or
clay, and let the horse stand in it when convenient, to
soften the hoofs. Spread the shoe a little every week
to help the hoofs out, or the shoes will prevent what

the frog pressure aims to do, but this spreading must be done with care. If the frog is dried up and hard, don't put on a bar shoe, as it will do more harm than good. In such a case make a common shoe with low or no calks; make holes in it as far back as you can nail; spread them with care a little every week. Let the horse stand in a box with mud or manure, even warm water, for a few hours at a time, and keep the hoofs moist with hoof ointment. In either case do not let the shoe stay on longer than four weeks at a time. In addition to the above pack the feet with some wet packing, or a sponge can be applied to the feet and held in position by some of the many inventions for this purpose.

No man can comprehend how much a horse suffers from contraction when his feet are hoof-bound and pressed together as if they were in a vise. The pain from a pair of hard and tight boots on a man are nothing compared to the agony endured by this noble and silent sufferer. It must be remembered that there is no such a thing as shoeing for contraction. Contraction is brought on by artificial living and shoeing. A bar shoe for contraction is the most foolish thing to imagine. The pressure intended on the frog is a dead pressure, and in a few days it will settle itself so that there is no pressure at all. If a bar shoe is to be used it must be an open bar shoe like the one referred to. This shoe will give a live pressure, and if made of stell will spring up against the frog at every step and it can be spread. I will say, however, that I don't recommend spreading, for it will part if not done with care. It is better to drive the shoe on with only four or five nails, and set them over often. Contraction never

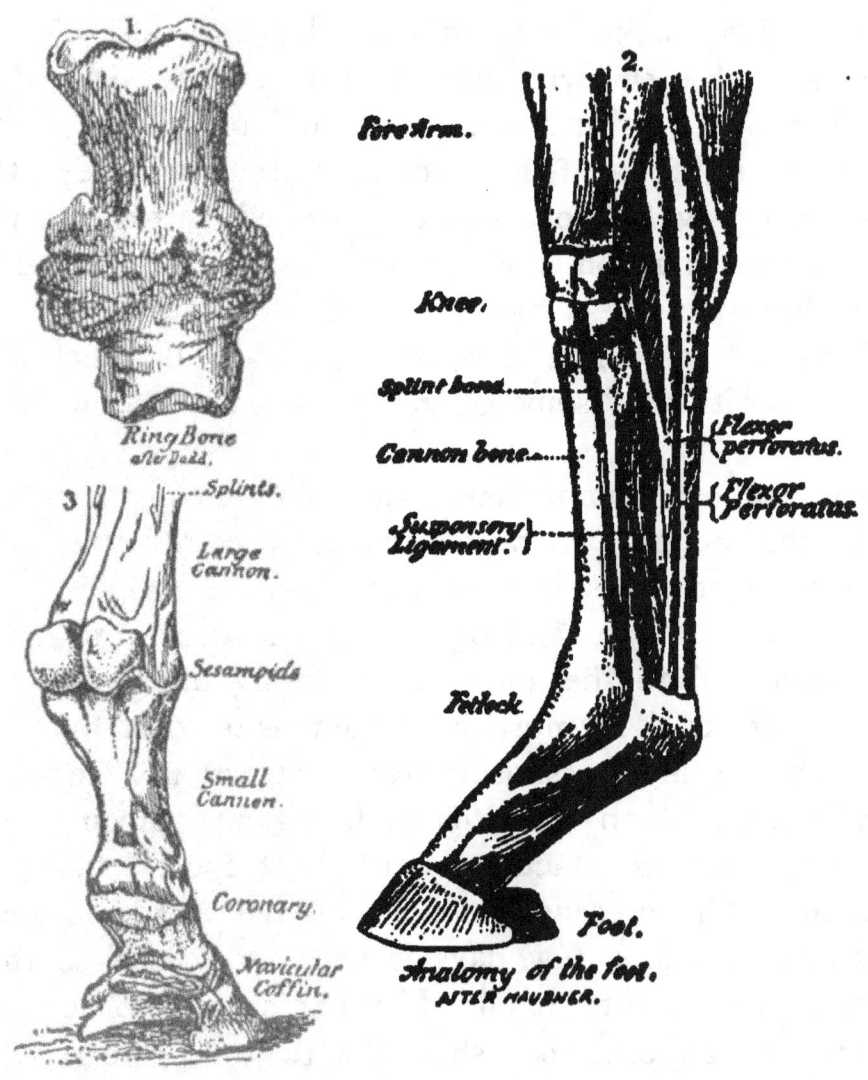

Anatomy of the feet.
AFTER MAUBNER.

affects the hind feet because of the moisture they receive. This should suggest to ever shoer that moisture is better than shoes.

CORNS

Corns are very common to horses' feet, a majority of all cases of lameness is due to this trouble.

Corns are the result of shoes being allowed to stay on too long. The shoe, in such a case grows under the foot and presses on the sole and corns are formed. Even pressure of the shoe and sometimes too heavy bearing on the heel causes corns. Gravel wedging in under the shoe or between the bar and the wall is sometimes the cause of corns. Leaving the heel and quarters too high, whereby they will bend under and press against the sole, is another cause of corns.

The seat of corns is generally in the sole of the foot at the quarter or heels between the bar and the wall, at the angle made by the wall and bar.

Anything that will bruise the underlying and sensitive membrane of the sole will produce corn. This bruise gives rise to soreness, the sole becomes blood colored and reddish; if bad it might break out, either at the bottom or the junction of the hoof and hair or coronet, forming a quittor.

Cut out the corn or red sole clear down. If the corn is the result of contraction pare down the hoof and sole, put the foot into linseed poultice that is warm, for twenty-four hours, then renew it. If the corn is deep, be sure to cut down enough to let the matter out. It is a good thing to pour into the hole hot pine tar. In shoeing the bearing should be taken off the quarter or from the wall over the corn by rasping it down so that it will not touch the shoe. A bar shoe is a good thing as it will not spring as much as to come in contact with the hoof over the corn. Give very little frog pressure. An open shoe can be used and in

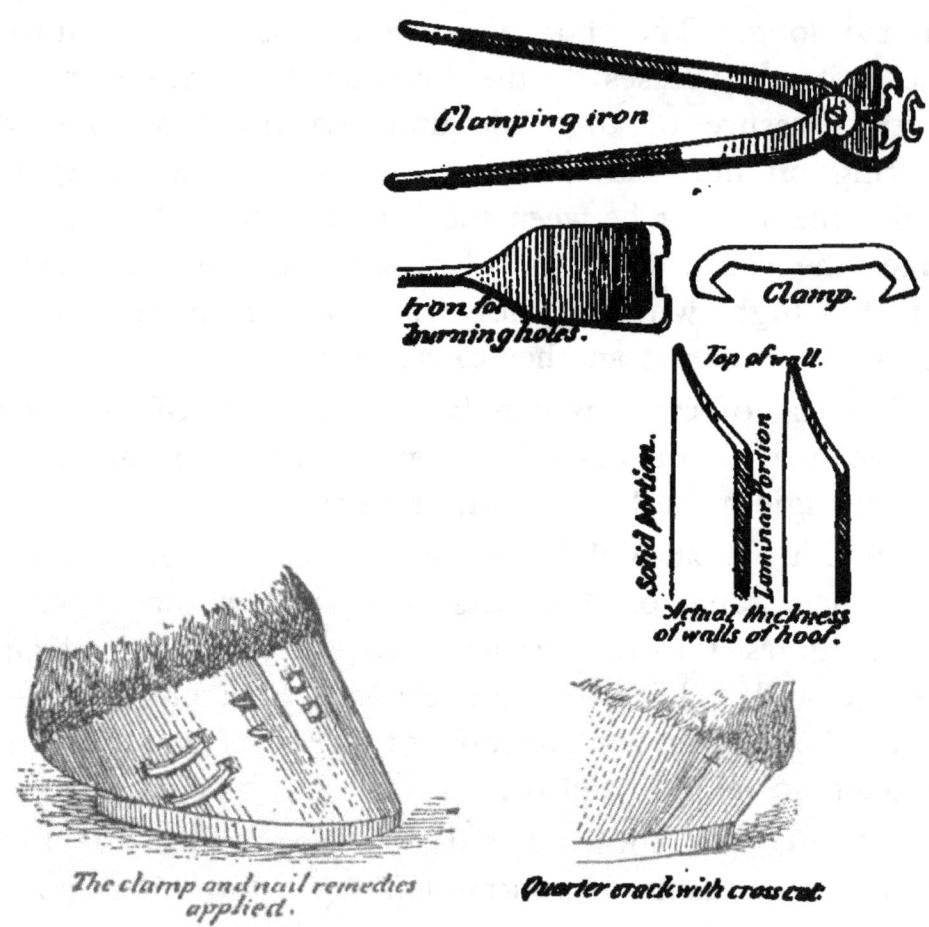

The clamp and nail remedies applied.

Quarter crack with cross cut.

such a case there should be no calk at the heel. A calk should be welded on directly over the corn and the shoe will not spring up against the wall.

QUARTER AND SAND CRACKS

Quarter and sand cracks are cracks in the hoof, usually running lengthwise of the fibers, but sometimes they will be running across the fiber for an inch

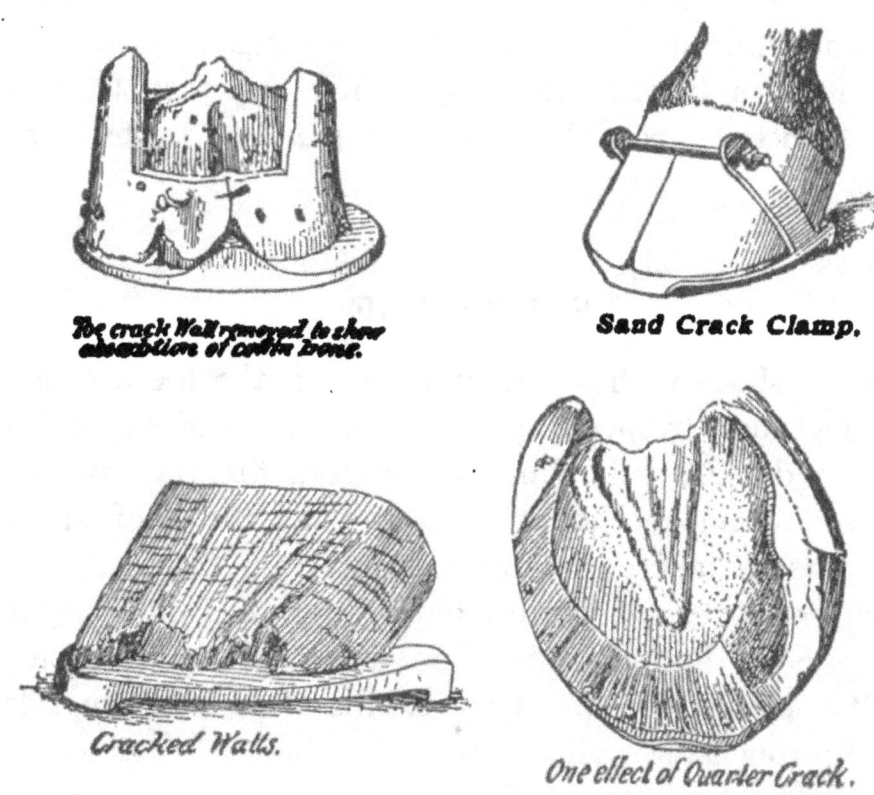

Toe crack Wall removed to show absorbtion of coffin bone.

Sand Crack Clamp.

Cracked Walls.

One effect of Quarter Crack.

or more. Quarter cracks are cracks mostly on the inside of the hoof, because that side is thinner and weaker than the outside. The cause of it is a hard and brittle hoof with no elasticity, brought on by poor assimilation and a want of good nutrition to the hoof. Hot, sandy or hard roads are also conducive to these cracks. What to do: If the horse is shod remove the shoes, and cut off the wall of the quarter to take off the bearing on both sides of the crack. If the crack goes up to the coronet and is deep, cut off both sides of the crack the whole length. About one inch below the coronet, cut a deep cut clear through either with a knife or hot sharp iron across the crack. This will help to start a new hoof.

If the flesh sticks up between the cracks, let a veterinarian burn it off. In shoeing for this trouble, it is best to use a bar shoe (endless) and shoe the horse often.

SEEDY TOE

When shoes with a clip or a cap on the toe are used it sometimes happens that the toe is bruised and it starts a dry rot extending up between the wall and the laminæ. Remove the shoe, pare away the hoof at the toe so as to take away the bearing from the toe. Any white or meaty substance should be picked out. Apply hot pine tar into the hole, and dip a little wad of tow in the hole to fill up. Replace the shoe, but don't let the clip touch the wall.

PRICKING

Pricking often happens in shoeing from a nail running into the quick, but the horse is often pricked by stepping on a nail or anything that will penetrate the sole and run into the quick. If the horse is pricked by shoeing pull off the shoes and examine each nail, the nail which has gone into the quick is wet and of a blue color.

If it is a bad case the sole or wall must be cut down to let the matter out and the foot put into a boot of linseed poultice. In milder cases a little pine tar put into the hole will be enough.

STIFLED

Mistakes are often made by inexperienced men and horse-shoers when a case of this kind is to be treated, and I would advise every horse-shoer to call in a veterinarian when he gets a case of this kind. Cramps of the muscles cf the thighs are sometimes taken for stifle.

When stifle appears in an old horse, three ounces of lead through his brain is the best, but for a young horse a cruel method of shoeing might be tried. Make a shoe with heels three inches high, or a shoe with cross bands as shown in illustration, Figure 8, No. 2, for stifle shoe. This shoe must be placed on the well foot. The idea is to have the horse stand on the stifled leg until the muscles and cords are relaxed.

STRING HALT

String halt or spring halt is a kind of affection of the hind legs, occasioning a sudden jerk of the legs upward towards the belly. Sometimes only one leg is affected.

In some cases it is milder, in others more severe. In some cases it is difficult to start the horse. He will jerk up on one leg and then on the other, but when started will go along all right.

For this fault there is no cure because it is a nervous affection. If there is any local disorder it is best to treat this, as it might alleviate the jerk. For the jerk itself bathe the hind quarters once a day with cold water. If this don't help try warm water, once a day for two weeks. Rub the quarters dry after bathing.

HOW TO SHOE A KICKING HORSE

Many devices are now gotten up for shoeing kicking horses. It is no use for a man to wrestle with a horse, and every horse-shoer should try to find out the best way to handle vicious horses.

One simple way, which will answer in most cases, is to put a twist on one of the horse's lips or on one ear. To make a twist, take a piece of broom handle two feet long, bore a half-inch hole in one end and put a piece of a clothes line through so as to make a loop six inches in diameter.

Another way: Make a leather strap with a ring in, put this strap around the foot of the horse; in the ring of the strap tie a rope. Now braid or tie a ring in the horse's tail and run the rope through this ring and back through the ring in the strap, then pull the foot up. See Fig. 16. The front foot can be held up by this device also, by simply buckling the strap to the foot and throwing the strap up over the neck of the horse.

Shoeing stalls are also used, but they are yet too expensive for small shops.

No horse-shoer should lose his temper in handling a nervous horse and abuse the animal; for, in nine cases out of ten, will hard treatment make the horse worse, and many horse owners would rather be hit themselves than to have anybody hit their horse.

EASY POSITION FOR FINISHING

Don't curse. Be cool, use a little patience, and you will, in most cases, succeed. To a nervous horse you should talk gently, as you would to a scared child. The horse is the noblest and most useful animal to man, but is often maltreated and abused. Amongst our dumb friends, the horse is the best, but few recognize this fact.

HOW TO SHOE A TROTTER

In shoeing a trotter it is no use to follow a certain rule for the angle, because the angle must vary a little in proportion to the different shape of the horse's foot.

Every owner of a trotter will test the speed by having shoes in different shapes and sizes, as well as having the feet trimmed at different angles, and when the angle is found that will give the best results the owner will keep a record of the same and give the horseshoer directions and points in each case.

The average weight of a horse-shoe should be eight ounces. Remember this is for a trotter. Make the shoe fit to the edges of the wall so that there will be no rasping done on the outside. In farm and draft horses this is impossible, as there is hardly a foot of such a uniform shape but what some has to be rasped off.

Use No. 4 nails, or No. 5.

Don't rasp under the clinches of the nails.

Make the shoes the shape of No. 1, Figure 8.

HOW TO SHOE A HORSE WITH POOR OR BRITTLE HOOFS

Sometimes it is difficult to shoe so as to make the shoe stay on on account of poor and brittle hoofs. In such a case the shoe should be fitted snug. Make a shoe with a toe clip.

HOW TO SHOE A WEAK-HEELED HORSE.

In weak heels the hoof is found to be low and thin from the quarters back. The balls are soft and tender. The shoes should not touch the hoof from the quarters back to the heels. An endless bar shoe is often the best thing for this trouble, giving some frog pressure to help relieve the pressure against the heels.

FOUNDER

Founder is a disease manifested by fever in the feet in different degrees from a simple congestion to a severe inflammation. It is mostly exhibited in the fore feet, being uncommon in the hind feet. The reason for this is the harder pressure, a much greater amount of weight coming on the front feet, the strain and pressure on the soft tissues heavier. The disease is either acute or chronic, in one foot or both. When both feet are diseased the horse will put both feet forward and rest upon the heels so as to relieve the pressure of the foot. If only one foot is affected that foot is put forward and sometimes kept in continual motion, indicating severe pain. The foot is hot, especially around the coronary band. The disease, if not checked, will render the horse useless. When such a horse is brought to you for shoeing it would be best to send him to a veterinarian.

How to shoe: Let the horse stand in a warm mud puddle for six hours, then put on rubber pads or common shoes with feet between the web of the shoe and the hoof, with sharp calks to take up the jar. It would be best not to shoe at all, but let the horse loose in a wet pasture for a good while.

"A righteous man regardeth the life of his beast."—Solomon.

CHAPTER XI

N this chapter the author desires to give some hints about the treatment for diseases most common to horses.

COLIC

There are two kinds of colic, spasmodic and flatulent.

Spasmodic colic is known by the pains and cramps being spasmodic, in which there are moments of relief and the horse is quiet.

Flatulent colic is known by bloating symptoms and the pain is continual, the horse kicks, paws, tries to roll and lie on his back.

For spasmodic colic give ½ ounce laudanum, ½ pint whisky, ½ pint water; mix well and give in one does. If this does not help, repeat the dose in half an hour.

For flatulent colic give ½ ounce laudanum, ½ ounce turpentine, ½ pint raw linseed oil, ¼ ounce chloroform, ½ pint water. Mix well and give in one dose. Repeat in one hour if the pain is not relieved.

BOTS

Sometimes there is no other symptom than the bots seen in the dung, and in most cases no other treatment is needed than some purgative.

MANGE

Mange is a disease of the skin due to a class of insects that burrow in the skin, producing a terrible itch and scab, the hair falling off in patches, and the horse rubs against everything. After the affected parts have been washed in soap-water quite warm, dry and rub in the following: 4 ounces oil of tar, 6 ounces sulphur, 1 pint linseed oil.

LICE

Make a strong tea of tobacco and wash the horse with it.

WORMS

There are many kinds of worms. Three kinds of tape worms and seven kinds of other worms have been found in the horse. The tape worms are very seldom found in a horse and the other kinds are easily treated by the following: One dram of calomel, 1 dram of tartar emetic, 1 dram of sulphate of iron, 3 drams of linseed meal. Mix and give in one dose for a few days; then give a purgative. Repeat in three weeks to get rid of the young worms left in the bowels in the form of eggs, but which have since hatched out.

DISTEMPER

Distemper is a disease of the blood. The symptoms are: Swelling under the jaws; inability to swallow, a mucous discharge from the nose.

Give the horse a dry and warm place and nourishing food. Apply hot linseed poultice to the swellings under the jaws and give small doses of cleansing powder for a few days.

HYDROPHOBIA

As soon as a case is satisfactorily recognized, kill the horse, as there is no remedy yet discovered that will cure this terrible disease.

SPAVIN

There are four kinds of spavin and it is difficult for any one but a veterinarian to tell one kind from another. In all cases of spavin (except blood spavin) the horse will start lame, but after he gets warmed up the lameness disappears and he goes all right until stopped and cooled off, when he starts worse than before.

There are many so-called spavin cures on the market, some of them good, others worse than nothing. If you don't want to call a veterinarian, I would advise you to use "Kendall's Spavin Cure." This cure is one of the best ever gotten up for this disease, and no bad results will follow the use of it if it does not cure. It is for sale by most druggists.

In nearly all cases of lameness in the hind leg the seat of the disease will be found to be in the hock-joint, although many persons (not having had experience) locate the difficulty in the hip, simply because they cannot detect any swelling of the hock-joint; but

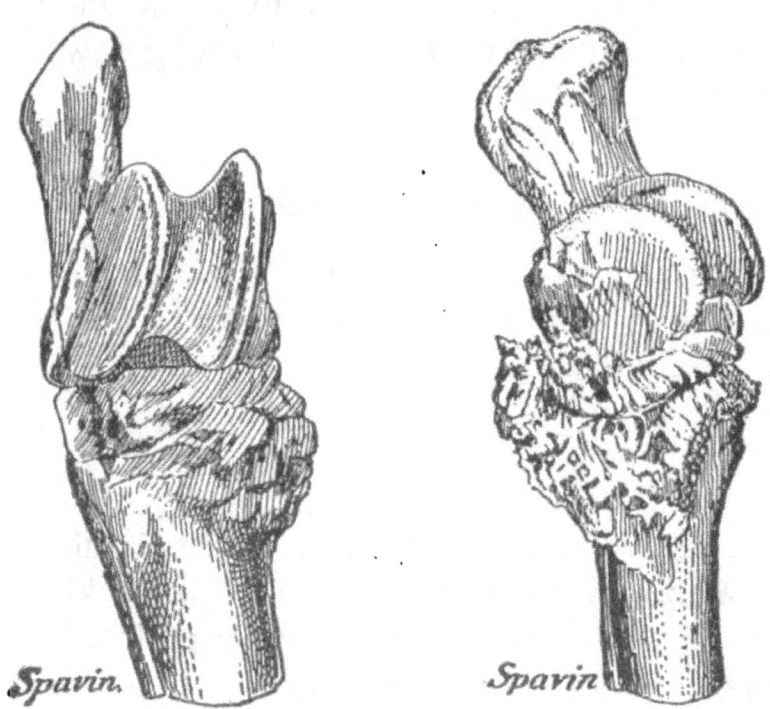

Spavin.   *Spavin*

in many of the worst cases there is not seen any swelling or enlargement for a long time, and perhaps never.

BONE SPAVIN

Bone spavin is a growth of irregular bony matter from the bones of the joint, and situated on the inside and in front of the joint.

Cause.—The causes of spavins are quite numerous,

but usually they are sprains, blows, hard work, and, in fact, any cause exciting inflammation of this part of the joint. Hereditary predisposition in horses is a frequent cause.

Symptoms.—The symptoms vary in different cases. In some horses the lameness comes on very gradually, while in others it comes on more rapidly. It is usually five to eight weeks before any enlargement appears. There is marked lameness when the horse starts out, but he usually gets over it after driving a short distance, and, if allowed to stand for awhile, will start lame again.

There is sometimes a reflected action, causing a little difference in the appearance over the hip joint, and if no enlargement has made its appearance, a person not having had experience is very liable to be deceived in regard to the true location of the difficulty. The horse will stand on either leg in resting in the stable, but when he is resting the lame leg he stands on the toe.

If the joint becomes consolidated the horse will be stiff in the leg, but may not have much pain.

Treatment.—That it may not be misunderstood in regard to what is meant by a cure, would say that to stop the lameness, and in most cases to remove the bunch on such cases as are not past any reasonable hopes of a cure.

But I do not mean to be understood that in a case of anchylosis (stiff joint), I can again restore the joint to its original condition; for this is an impossibility, owing to the union of the two bones, making them as one. Neither do I mean that, in any ordinary case of bone spavin which has become completely ossified (that is, the bunch become solid bone), that, in such a case, the enlargement will be removed.

In any bony growths, like spavin or ringbone, it will be exceedingly difficult to determine just when there is a sufficient deposit of phosphate of lime so that it is completely ossified, for the reason that in some cases

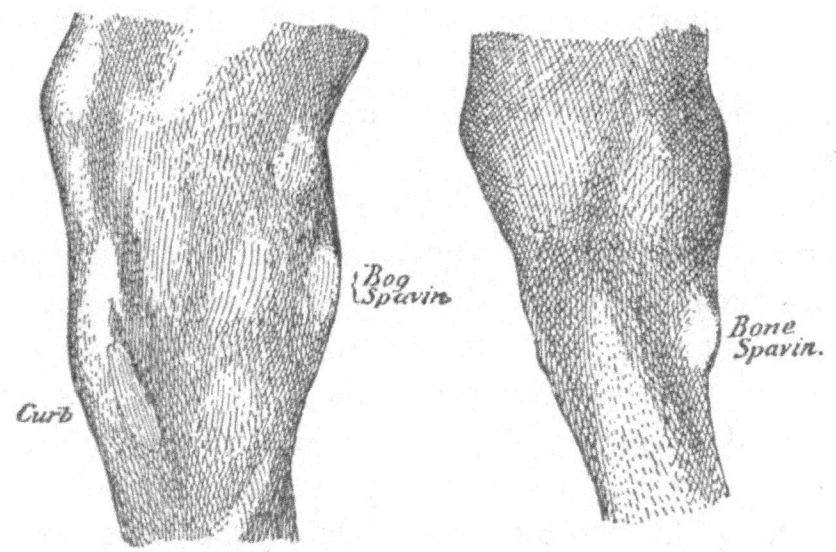

the lime is deposited faster than in others, and therefore one case may be completely ossified in a few months, while in another it will be as many years.

The cases which are not completely ossified are those that I claim to remove. One of this class which I have seen removed was a large bone spavin of four or five years standing, and I think that a large per cent of cases are not fully ossified for several months or years.

I am well aware that many good horsemen say that it is impossible to cure spavins, and, in fact, this has been the experience of horsemen until the discovery of Kendall's Spavin Cure. It is now known that the treatment which we recommend here will cure nearly every case of bone spavin which is not past any reasonable hopes of a cure, if the directions are followed, and the horse is properly used.

OCCULT SPAVIN

This is similar to bone spavin in its nature, the difference being that the location is within the joint, so that no enlargement is seen, which makes it more difficult to come to a definite conclusion as to its location, and consequently the horse is oftentimes blistered and tormented in nearly all parts of the leg but in the right place.

The causes and effects are the same as in bone spavin, and it should be treated in the same way.

These cases are often mistaken for hip disease, because no enlargement can be seen.

BOG SPAVIN

The location of this kind of a spavin is more in front of the hock-joint than that of bone spavin, and it is a soft and yet firm swelling. It does not generally cause lameness.

BLOOD SPAVIN

This is similar to bog spavin but more extended, and generally involves the front, inside and outside of the joint, giving it a rounded appearance. The swelling is soft and fluctuating.   Young horses and colts, especially if driven or worked hard, are more liable to have this form of spavin than older horses.

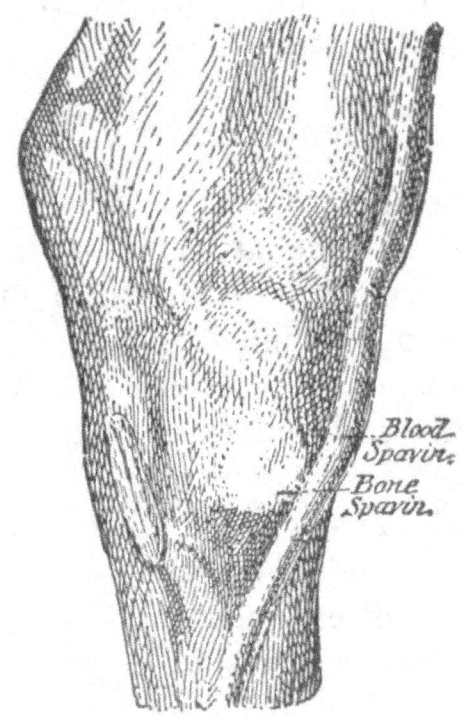

SPLINT

This is a small, bony enlargement, and generally situated on the inside of the foreleg about three or four inches below the knee joint, and occurs frequently in young horses when they are worked too hard.

SPRAIN

By this is meant the sudden shifting of a joint farther than is natural, but not so as to produce dislocation.

Every joint is liable to sprain by the horse's falling, slipping, or being overworked. These cases cause a great deal of trouble, oftentimes producing lameness, pain, swelling, tenderness, and an unusual amount of heat in the part.

Treatment.—Entire rest should be given the horse, and if the part is found hot, as is usually the case, apply cold water cloths, changing frequently, for from one to three days until the heat has subsided, when apply Kendall's Spavin Cure, twice or three times a day, rubbing well with the hand.

If the fever is considerable, it might be well to give fifteen drops of tincture of aconite root, three times a day, for one or two days, while the cold water cloths are being applied. Allow the horse a rest of a few weeks, especially in bad cases, as it is very difficult to cure some of these cases, unless the horse is allowed to rest.

STAGGERS

A disease of horses, resulting from some lesion of the brain, which causes a loss of control of voluntary motion. As it generally occurs in fat horses which are well fed, those subject to these attacks should not be overfed. The cause is an undue amount of blood flowing to the brain.

Treatment.—The aim of the treatment should be to remove the cause. In ordinary cases give half a pound of epsom salts, and repeat if necessary to have it physic, and be careful about overfeeding.

In mad staggers, it would be well to bleed from the neck in addition to giving the epsom salts.

CERTAIN CURE FOR HOG CHOLERA

Take the following ingredients well mixed together, and give one tablespoonful daily in food during sickness, and as a preventative two or three times a week:

Powdered charcoal ....1 pound
"   mandrake 2   "
"   resin 1   "
"   saltpeter ....8 ounces
"   madder 8   "
"   bi-carbonate of soda.......6 pounds

TENSILE STRENGTH OF IRON AND OTHER MATERIALS

Pounds required to tear asunder a rod one inch square:

Cast steel 145,000
Soft steel 115,000
Swedish iron 85,000
American iron 60,000
Russian iron 62,000
Wrought wire 98,000
Cast iron, best....................... 45,000
Cast iron, poor....................... 14,000
Silver 40,000
Gold 21,000
Whalebone 8,000

Bone............................	8,000
Tin.............................	5,000
Zinc...........................	3,000
Platinum.......................	40,000
Boiler plates...................	50,000
Leather belt (lin.)...............	350
Rope (manila)	10,000
Hemp (tarred).................	14,000
Brass	40,000

HOW CORN IN THE CRIB AND HAY IN THE MOW SHOULD BE MEASURED

As near as can be figured out, two cubic feet of corn in the ear will make one bushel shelled. To find the quantity of corn in the crib, measure length, breadth and height, multiply the breadth by the length and this product by the height; then divide this product by two, and you have the right number of bushels of corn.

It is estimated that 510 cubic feet of hay in a mow will make one ton. Multiply the length by the breadth and the product by the height; divide this product by 510, and the quotient shows the tons of hay in the mow.

GRAIN SHRINKAGE

Not often do the farmers gain any by keeping the grain, for it will shrink more than the price will make good. Wheat will shrink 7 per cent in seven months from the time is is thrashed. Therefore, 93 cents a bushel for wheat in September is better than $1 in

April the following year. Add to this the interest for the money you could have used in paying debts, or loaned, and it will add 4 per cent more, making it 11 per cent.

Corn will shrink more than wheat, and potatoes are very risky to keep on account of the diseases they are subjected to; the loss is estimated at 30 per cent for six months.

VALUE OF A TON OF GOLD OR SILVER

A ton of gold is worth in money $602,799.21; a ton of silver, $37,704.84.

AGES OF ANIMALS

	Years.
Elephant	1 to 400
Whale	100
Swan	250
Eagle	100
Raven	110
Stag	50
Lion	75
Mule	75
Horse	30
Ox	30
Goose	75
Hawk	35
Crane	24
Skylark	20
Crocodile	100
Tortoise	150
Cow	20

Deer.. 20
Wolf....................................... 20
Swine...................................... 20
Dog.. 12
Hare 8
Squirrel 7
Titlark.................................... 5
Queen bee 4
Working bee.....................6 months

RINGWORM

Ringworm is a contagious disease and attacks all kinds of animals, but it often arises from poverty and filth. It first appears in a round bald spot, the scurf coming off in scales.

Cure: Wash with soap-water and dry, then apply the following once a day. Mix 25 grains of corrosive sublimate in half a pint of water and wash once a day till cured.

BALKING

Balking is the result of abuse. If a horse is overloaded and then whipped unmercifully to make the victim perform impossibilities, he will resent the abuse by balking.

There are many cruel methods for curing balking horses, but kindness is the best. Don't hitch him to a load he cannot easily pull. Let the man that is used to handling him drive him. Try to divert his mind from himself. Talk to him; pat him; give him a handful of oats or salt. But if there is no time to wait pass a chain or rope around his neck and pull him along with another horse. This done once all there is needed, in most cases, is to pass the rope around and the horse will start. It is no use trying to whip a balking horse, because balking horses are generally horses of more than common spirit and determination, and they will resent abuse every time. Kindness, patience and perseverance are the best remedies.

RATTLE-SNAKE BITE

When a horse has been bitten by a rattlesnake, copperhead, or other venomous serpent, give the following: One-half teaspoonful of hartshorn, 1 pint whisky, ½ pint of warm water. Mix well and give one dose. Repeat in one hour if not relieved. Burn the wound at once with a hot iron, and keep a sponge soaked in ammonia over the wound for a couple of hours.

HOOF OINTMENT

Rosin, 4 ounces; bees wax, 4 ounces; pine tar, 4 ounces; fish oil, 4 ounces; mutton tallow, 4 ounces. Mix and apply once a day.

PURGATIVE

Aloes, 3 drams; gamboge, 2 drams; ginger, 1 dram; gentian, 1 dram; molasses, enough to combine the above. Give in one dose, prepared in the form of a ball.

HINTS TO BLACKSMITHS AND HORSE-SHOERS

Don't burn the shoe on.

Don't rasp under the clinchers.

Don't rasp on the outer side of the wall more than is absolutely necessary.

Don't rasp or file the clinch heads.

Don't make the shoes too short. Don't make high calks. Don't pare the frog.

Don't cut down the bars. Don't load the horse down with iron.

Don't lose your temper. Don't hit the horse with the hammer.

Don't run down your competitor. Don't continually tell how smart you are.

Don't smoke while shoeing. Don't imbibe in the shop. Don't run outdoors while sweaty. Don't know it all. Always be punctual in attendance to your business. Allow your customers to know something. No man is such a great fool but that something can be learned of him.

Be always polite. Keep posted on everything belonging to your trade. Read much. Drink little. Take a bath once a week. Dress well. This done, the craft will be elevated, and the man respected.

ADVICE TO HORSE OWNERS

T is cruelty to animals to raise a colt and not train him for shoeing, and t h e horse - shoer must suffer for this neglect also. Many a valuable horse has been crippled or maltreated, and thousands of horse - shoers suffer hardships, and many are crippled, and a few killed every year for the horse owner's carelessness in this matter. A law should be enacted making the owner of an ill-bred horse responsible for the damage done to the horse-shoer by such an animal. Every horse-raiser should begin while the colt is only a few days old to drill him for the shoeing. The feet should be taken, one after the other, and held in the same position as a horse-shoer does, a light hammer or even the fist will do, to tap on the foot with, and the feet should be handled and manipulated in the same manner the horse-shoer does when shoeing. This practice should be kept up and repeated at least once a week and the colt when brought to the shop for shoeing will suffer no inconvenience. The horse-shoer's temper, as well as muscles, will be spared and a good feeling all around prevails.

Horse-raisers, remember this.

ADVICE TO YOUNG MEN

In every profession and trade it is a common thing to hear beginners say: I know, I know. No matter what you tell them, they will always answer, I know. Such an answer is never given by an old, learned or experienced man, because, as we grow older and wiser we know that there is no such thing as knowing it all. Besides this we know that there might be a better way than the way we have learned of doing the work. It is only in few cases that we can say that this is the best way, therefore we should never say, I know: first, because no young man ever had an experience wide enough to cover the whole thing; second, it is neither sensible nor polite. Better not say anything, but simply do what you have been told to do.

Every young man thinks, of course, that he has learned from the best men. This is selfish and foolish. You may have learned from the biggest botch in the country. Besides this, no matter how clever your master was, there will be things that somebody else has a better way of doing. I have heard an old good blacksmith say, that he had never had a helper but what he learned some good points from him.

Don't think it is a shame, or anything against you, to learn. We will all learn as long as we live, unless

we are fools, because fools learn very little.   Better to
assume less than you know than to assume more.

Thousands of journeymen go idle because many a
master would rather hire a greenhorn than hire a
"knowing-it-all" fellow.   Don't make yourself obnoxious by always telling how your boss used to do this or
that.   You may have learned it in the best way possible, but you may also have learned it in the most
awkward way.   First find our what your master
wants, then do it, remembering there are sometimes
many ways to accomplish the same thing.   Don't be
stubborn.   Many mechanics are so stubborn that they
will never change their ways of doing things, nor
improve on either tools or ideas.

Don't be a one-idea man; and remember the maxim,
"A wise man changes his mind, a fool never."

Be always punctual, have the same interest in doing
good work and in drawing customers as you would
were the business yours.   Be always polite to the
customers, no matter what happens.   Never lose your
temper or use profane language.   Don't tell your
master's competitors his way of doing business, or
what is going on in his dealings with people.   You are
taking his money for your service, serve as you would
be served.

IRON CEMENT

A cement for stopping clefts or fissure of iron vessels
can be made of the following: Two ounces muriate
of ammonia, 1 ounce of flowers of sulphur, and 1

pound of cast-iron filings or borings. Mix these well in a mortar, but keep the mortar dry. When the cement is wanted, take one part of this and twenty parts of clean iron borings, grind together in a mortar. Mix water to make a dough of proper consistence and apply between the cracks. This will be useful for flanges or joints of pipes and doors of steam engines.

HOW TO RUN A TURNING LATHE

(By a student of James College of Mechanic Arts, at Ames, Iowa.)

Lathes, when first invented, were very rude affairs, but they, like all other machinery, have experienced improvement from year to year until now some of them are more complicated than a watch, and for that reason should receive the best of care. They should be kept clean and well oiled. While being used the dust and shavings should be cleaned off at least every night, and every half day is better.

When they are kept in a dusty place, as is very often the case in a general repair shop, they should be kept covered while not in use. Some cheap canvas makes a good cover.

Every person who intends running a lathe should first become acquainted with his machine; become familiar with all the combinations that can be made, so that when a piece of work comes in to be done he will know just how to arrange the lathe to do that work. For instance, a piece of work needs to be turned tapering; this is done by shifting the tail stock to one side. Or there are threads to be cut; know just how to arrange the lathe to cut any number of threads to the inch.

Next to care of lathe comes care of tools. When there are a few minutes spare time see that the tools

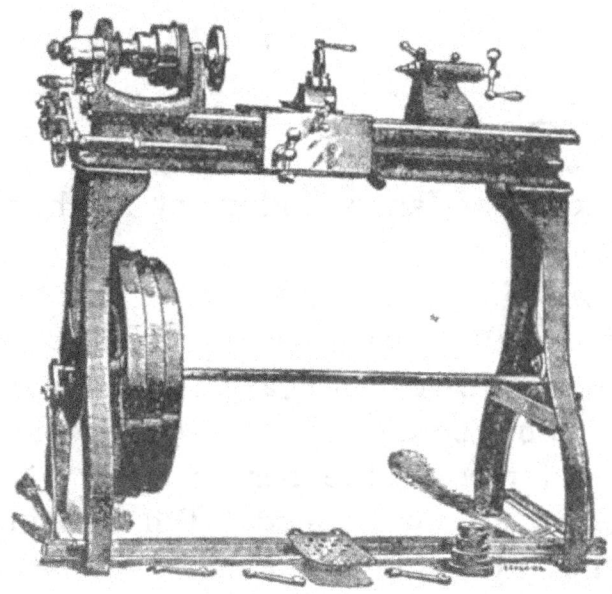

are sharp. Keep them sharp. They will do the work better, faster and with much less strain on the machine.

All cutting tools should be made diamond shape, with either one side or the other, depending on the way the carrier is to move, made a little higher; the right side being highest when the carrier is moving to the right, and vice versa. The sharp edge of smoothing tools is made square across, like a plane bit, and thread-cutting tools should be made the same shape as the thread to be cut.

Water or oil should be kept on the iron or steel that is being turned. It keeps the point of the tool from getting hot when heavy chips are taken, and it makes a smoother job when the smoothing tool is used. There is no need to use either water or oil when turning cast iron.

The tempering of lathe tools is a very particular piece of work, varying considerable with the kind of steel used and the nature of the work to be done. For slow heavy turning the tool must not be too hard, else it would break; while for light swift turning it should be quite hard. For water tempering the temper color varies from a dark blue to a very light straw color, depending, as I have said before, on the nature of the work to be done.

By way of illustration of a piece of work that represents a number of lathe combinations, I will take the fitting of a saw shaft for our common wood saws. First place the balance wheel in the lathe chuck, being sure to get it in the center, so that when the hole is drilled in the wheel it will be in the exact center. Take a drill a sixteenth of an inch smaller than the hole to be made, and drill out the hole. Use the inside boring tool to make the hole the desired size. Turn a smooth face on the hub of the wheel where it comes against the box; then the wheel is ready for the key seat. To cut the key seat in the wheel use a key-seat chisel the same size as the milling wheel used to cut the key seat in the shaft.

Next take one of the saw collars; put it in the chuck, being careful to get this in the center also, with the widest side next the chuck, and drill a hole in it the same size as the hole in the saw. Turn off the end of the collar to get it square. Prepare the other collar in the same way.

Now cut the shaft off the length wanted, and turn one end to fit tightly into the balance wheel. Turn off a place next to where the wheel comes for the bearing or box. Now turn the shaft around and fit the other end for the collars. The collar that goes on the inside or side next the bearing should be shrunk on. To do this leave the shaft about one sixty-fourth of an inch larger than the hole in the collar, then heat the collar to a red heat, and slip it onto the shaft. It should not be driven very hard, or it will break in cooling. Let it cool of its own accord. When nearly cool it can be put into water and cooled off.

The next step is to true up the inside of the collar, leaving about one inch of surface to come against the saw. Now turn the shaft down to the size wanted for the thread, either 1-inch or 1⅛-inch, then with a cut-off tool about ⅛-inch wide, cut in next the shoulder the depth of the thread. If there is a die and tap handy that will be the quickest way to cut the thread, but if not handy then use the lathe. Now screw the nut on and turn off the inside of the nut. For fitting the loose collar there should be on hand a shaft about 14 or 16 inches long, turned a very little tapering; then drive the collar onto this shaft and finish it up. When ready put this collar into place on the saw shaft and screw the nut up tight. Now smooth off the outside of the collars for loops. Cut the key seat in the shaft and key the balance wheel on solid, being careful to get the distance between the wheel and the saw collar the exact distance between the outside of the boxes.

HOW TO BALANCE A PULLEY

When a pulley or balance wheel is to be balanced you must first have a shaft that is of the same size as the hole in the pulley. Of course, the wheel or pulley must be turned and trued up so that it is finished before you balance the same.

After the shaft has been put in and tightened, place two pieces of angle iron or T-iron about two feet long parallel on a pair of wooden horses. The irons must be level. Now place the pulley between the irons so that the shaft will have a chance to roll on the "T" or angle iron, and you will notice that the heaviest side of the pulley will be down. Start it rolling, and the pulley will always stop with the heaviest side down. Now, if the pulley or wheel, as the case may be, has a thick rim, then bore out from the heaviest side enough to balance, or you can drill a hole in the lightest side and bolt a piece of iron to it just heavy enough to balance the wheel.

HOW TO PUT IN A WOODEN AXLE

One of the most difficult pieces of work to do in a wagon shop is to put in a wooden axle.

In the first place, you must have well-seasoned timber, hickory or maple. Take out the old axle. The skeins will come off easy by heating them a little. Now cut the timber the exact length of the broken axle. In order to get the right pitch and gather, you must cut off one-half inch from the back side of the end of the timber and one-half inch from the bottom side, this cut to run out at the inner end or collar of the skein, as shown in Figure 14. Next take dividers and make a circle in the end of the axle the size of the old axle—in case new skein is put on, the size of the

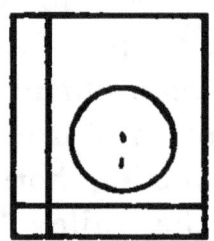

bottom of the skein inside. This circle must be made so that the lower side of it will go down to edge of the timber, and the sides be of the same distance from the edges. You will now notice that most of the hewing will be done on top side, as it must in order to get the right pitch, and as one-half inch has been cut from the back side it will throw the front side of the wheel in a little; this is gather. If a wheel has no gather the wheel will be spread out against the nut of the skein, and the wear will be in that direction, and the wheel will rattle, as you know the skein is tapered; but if the wheel has gather, the pressure will be against the collar of the skein, and the wheel will be tight, as it forces itself up against the collar and the wider end of the skein.

Some wagon-makers will use the old axle as a guide and cut the new by the old. This is not safe, as the old is mostly sprung out of shape.

In hewing the axle for the skein great care should be taken not to cut off too much; better go slow, because it depends upon the fitting of the skein to get a good job. When the axle is finished or ready to be driven into the skein be sure to have the axle strong; that is, a little too large to go in easy. Now warm—or heat, if you will—the skein a little, not so much that it will burn, and drive it onto its place by a mallet. In making new wagons I think it would be wise to paint the part of the axle that goes in the skein, but in repairing I deem it unwise, because it will have a tendency to work loose unless it will have time to dry before using, and I have noticed paint to be still fresh in the skein after years of use. There should be no gap left between the collar of the skein and the axle, as water will run in and rot the timber.

HOW TO PUT IN SPOKES

VERY wagon-maker is supposed to know how to put in spokes. Still, there are sometimes wagon - makers, especially beginners, that don't know. First clean out the sliver left of the old spoke, and make the mortise dry, and in every case use glue. In a buggy wheel take the rivet or rivets out, if there is any, and be sure to have the right shape of the tenon to fit the mortise in the hub, so as to make the spoke stand plumb. Set the tenon going through the rim. Be sure to have this tenon reach through. This is important in filling a wagon wheel, because, if the tenons don't reach through the fellow, then the heft will rest against the shoulder of the tenon, and when the tire is put on tight and the wagon used in wet roads, the fellow will soften and the spokes settle into the rim. The tire gets loose, and some one, either the wagon-maker or the blacksmith, will be blamed—in most cases the blacksmith. Of course, the tenon should not be above the rim. After the spokes have been put in rivet the flange of the hub, or so many rivets as you have taken out. This should always be done before the tire is set.

WEIGHT OF ONE FOOT IN LENGTH OF SQUARE AND ROUND BAR IRON

Size.	Square.	Round.	Size.	Square.	Round.
1/4	.209	.164	2 1/8	15.000	11.840
5/16	.326	.256	2 1/4	16.900	13.280
3/8	.469	.368	2 3/8	18.835	14.792
7/16	.638	.504	2 1/2	20.871	16.392
1/2	.833	.654	2 5/8	23.112	18.142
9/16	1.057	.831	2 3/4	25.250	19.840
5/8	1.305	1.025	2 7/8	27.600	21.681
11/16	1.579	1.241	3	30.065	23.650
3/4	1.875	1.473	3 1/8	32.610	25.615
13/16	2.201	1.728	3 1/4	35.270	27.702
7/8	2.552	2.004	3 3/8	38.040	29.875
15/16	2.930	2.301	3 1/2	40.900	32.160
1	3.340	2.625	3 5/8	43.860	34.470
1 1/8	4.222	3.320	3 3/4	46.960	36.890
1 1/4	5.215	4.098	3 7/8	50.150	39.390
1 3/8	6.310	4.960	4	53.435	41.980
1 1/2	7.508	5.900	4 1/4	60.320	47.380
1 5/8	8.810	6.920	4 1/2	67.635	53.130
1 3/4	10.200	8.040	4 3/4	75.350	59.185
1 7/8	11.740	9.222	5	83.505	65.585
2	13.300	10.490	6	120.240	94.608

WEIGHTS OF ONE LINEAL FOOT OF FLAT BAR IRON

Thickness.	Width, 1.	Width, 1¼.	Width, 1½.	Width, 1¾.
⅛	.416	.521	.624	.728
3⁄16	.625	.780	.938	1.090
¼	.833	1.040	1.250	1.461
5⁄16	1.041	1.301	1.560	1.821
⅜	1.252	1.562	1.881	2 190
7⁄16	1.462	1.822	2.191	2.550
½	1.675	2.085	2.505	2.925
9⁄16	1.884	2.345	2.815	3.285
⅝	2.085	2.605	3.132	3 655
11⁄16	2.295	2 860	3.442	4 010
¾	2.502	3.131	3.752	4 381
⅞	2 921	3.650	4 382	5 100
1	3.331	4.170	5 005	5.832
1⅛	3.750	4.694	5.630	6.560
1¼	4.175	5.210	6.251	7.290
1⅜	4.580	5.728	6.879	8.022
1½	5.005	6.248	7 502	8.750
1⅝	5.425	6 769	8 130	9.480
1¾	5.832	7.289	8.749	10.208
1⅞	6 248	7.800	9 380	10.938
2	6.675	8 332	10.005	11.675

WEIGHTS OF ONE LINEAL FOOT OF FLAT BAR IRON

(Continued)

Thick-ness.	Width, 2.	Width, 2¼.	Width, 2½.	Width, 2¾.
⅛	.832	9.370	1.040	1.151
3/16	1.251	1.410	1.562	1.720
¼	1.675	1 878	2.080	2.290
5/16	2.081	2.342	2.000	2.862
⅜	2.502	2.811	3.135	3.445
7/16	2 920	3 278	3.650	4.010
½	3.335	3.748	4.175	4.580
9/16	3 748	4.220	4.089	5 160
⅝	4.168	4.690	5.211	5 730
11/16	4.578	5.160	5.735	6.150
¾	5.005	5.630	6.255	6.880
⅞	5.830	6.558	7.395	8.025
I	6.668	7.500	8.332	9.170
1 ⅛	7 498	8.441	9.382	10 310
1 ¼	8.333	9.382	10.421	11.460
1 ⅜	9.775	10.310	11.460	12.605
1 ½	10.000	11.255	12.505	13.750
1 ⅝	10 835	12.190	13.545	14 905
1 ¾	11.675	13.135	14.585	16.045
1 ⅞	12.505	14 065	15.635	17.195
2	13.335	15.000	16.675	18.335

WEIGHTS OF ONE LINEAL FOOT OF FLAT BAR IRON

(*Continued*)

Thick-ness.	Width, 3.	Width, 3¼.	Width, 3½.	Width, 3¾.
⅛	1.250	1.350	1.465	1.658
3/16	1.879	2.035	2.195	2.345
¼	2.505	2.710	2.925	3.135
5/16	3.135	3.391	3.650	3.901
⅜	3.750	4.060	4.380	4.695
7/16	4.385	4.740	5.105	5.470
½	5.000	5.425	5.832	6.250
9/16	5.635	6.090	6.565	7.030
⅝	6.255	6.775	7.290	7.805
11/16	6.885	7.455	8.020	8.590
¾	7.500	8.135	8.750	9.380
⅞	8.750	9.480	10.210	10.940
1	10.000	10.835	11.675	12.500
1⅛	11.255	12.190	13.135	14.065
1¼	12.505	13.540	14.585	15.635
1⅜	13.750	14.905	16.045	17.195
1½	15.000	16.250	17.500	18.750
1⅝	16.255	17.605	18.960	20.310
1¾	17.505	18.965	20.425	21.880
1⅞	18.750	20.305	21.885	23.445
2	20.000	21.670	23.335	25.000

WEIGHTS OF ONE LINEAL FOOT OF FLAT BAR IRON

(Continued)

Thickness.	Width, 4.	Width, 4¼.	Width, 4½.	Width, 4¾.
⅛	1.670	1.774	1.887	1.989
3/16	2.500	2.658	2.811	2.971
¼	3.331	3.538	3.750	3.960
5/16	4.168	4.430	4.689	4.950
⅜	5.000	5.311	5.630	5.940
7/16	5.831	6.200	6.560	6.930
½	6.670	7.082	7.502	7.925
9/16	7.500	7.965	8.435	8.910
⅝	8.330	8.855	9.380	9.900
11/16	9.165	9.740	10.310	10.890
¾	10.000	10.630	11.250	11.880
⅞	11.670	12.400	13.140	13.845
1	13.340	14.165	15.000	15.830
1⅛	15.000	15.940	16.880	17.815
1¼	16.660	17.710	18.755	19.179
1⅜	18.335	19.480	20.650	21 770
1½	20.000	21.255	22.505	23.750
1⅝	21.675	23.025	24.380	25.730
1¾	23.335	24.790	26.240	27.710
1⅞	25.000	26.560	28.140	29.000
2	26.670	28.335	30.000	31.670

INDEX.